FASHIONING
FABRIC

THE ARTS OF SPINNING AND WEAVING IN EARLY CANADA

FASHIONING FABRIC

THE ARTS OF SPINNING AND WEAVING IN EARLY CANADA

ADRIENNE D. HOOD

JAMES LORIMER & COMPANY LTD., PUBLISHERS
TORONTO

James Lorimer & Company Ltd., Publishers acknowledges the support of the Ontario Arts Council. We acknowledge the support of the Government of Canada through the Book Publishing Industry Development Program (BPIDP) for our publishing activities. We acknowledge the support of the Canada Council for the Arts for our publishing program. We acknowledge the support of the Government of Ontario through the Ontario Media Development Corporation's Ontario Book Initiative.

Library and Archives Canada Cataloguing in Publication

Hood, Adrienne Dora, 1949–
    Fashioning fabric: the arts of spinning and weaving in early Canada / Adrienne D. Hood.

Includes bibliographical references and index.
ISBN-13: 978-1-55028-980-0
ISBN-10: 1-55028-980-2

    1. Hand spinning — Canada — History. 2. Hand weaving — Canada — History.
    3. Textile fabrics — Canada — History — 19th century. I. Title.

TT847.H65 2007   746.10971   C2007-900336-2

Illustration opposite title page: *Boutonné* coverlet from Quebec. Illustration on title page: Combed flax

James Lorimer & Company Ltd., Publishers
317 Adelaide Street West, Suite #1002
Toronto, ON
M5V 1P9
www.lorimer.ca

Printed and bound in Canada

Contents

Introduction

Every spring in early Canada, sheep dotted the countryside, and in summer, the blue of flax flowers waved in the wind. By fall, they would both provide raw material for the production of wool and linen, which would be made into warm clothing and essential household textiles such as bedding.

Although many think of the spinning wheel and the loom as central to their images of the past, cloth-making, as we know it, was not indigenous to North America. While there is archaeological evidence to suggest that Native peoples used twining and plaiting techniques to create containers, and constructed clothing using reeds, grasses, bark, and animal hair, very early on in Native-European relations, cloth manufactured in Europe became an important trade good.

As early as 1000 B.C.E. the Vikings, the very first Europeans to establish settlements in North America, left material proof of textile activity. Arriving in Newfoundland at what is today called L'Anse aux Meadows, Scandinavian men and women, who came to obtain lumber and fish, left behind remnants of tools that

Above: Canada's Native people used reeds, grasses, bark and animal hair to make clothing and baskets. European textile materials and tools were unfamiliar to indigenous peoples.
Facing page: In spite of their warm wool, sheep had to be sheltered during Canada's cold winters. Every spring, farmers sheared off the heavy wool that would later be spun and woven into warm clothing.

The flowers of the flax plant develop into seed pods. Removing the seeds in a process called "rippling" was one of the many procedures required to obtain flax fibre from the long stems.

A drop spindle, like this, is one of the simplest ways to make yarn. A weight called a "whorl" facilitates the rotation required to spin the yarn. This type of spindle could be used indoors or out, while standing or walking.

indicate they spun yarn and made knit goods. The spindle whorl, which would have been attached to a wooden shaft and used to twist fibre into yarn, and the knitting needle found in recent archaeological excavations make sense in what was an impermanent settlement — spinning with a spindle is a portable activity and can even be done while walking, as can knitting. The warm garments knit by the Viking women while they were camped in Newfoundland would have supplemented the clothing they brought with them until they returned home.

It was later European settlers, however, who introduced sheep and fibre-producing plants like hemp and flax to the New World. Along with fibre, artisans from Europe brought the technology — including the spinning wheels and looms — that was required to turn raw material into cloth.

Over the course of the eighteenth and nineteenth centuries, cloth-making developed, first in homes and later in factories. Although in larger, more metropolitan centres, imported cloth continued to be available, throughout that period, in rural Canada, prior to the craft revivals of the twentieth century, fibre production, hand spinning, and weaving filled many of the needs of individual households or supplied small local markets. While some urban women may have practised textile crafts, such as knitting, quilt-making, and fancy needlework, they would not have engaged in spinning and weaving.

The tools involved in making a piece of cloth were numerous and the work was labour-intensive. Producing the actual fibre was just the beginning. The raw material had to be prepared for spinning before it could be twisted into yarn. The yarn then had to be somehow interlocked to form a piece of cloth, for example, by felting, knitting, knotting, crocheting, or weaving. Finally that cloth had to be further processed, perhaps by dyeing, felting, or bleaching. Until the introduction, in the nineteenth century, of carding and spinning mills, and later of woollen and cotton mills that manufactured a finished product, the basic technology required to produce yarn or to weave cloth did not change.

Decoration, however, was another matter. Eastern Canada was settled by people from a variety of European cultures, including the French, the British, and the Germans, and each group ornamented and designed their textile tools in unique ways to suit their own aesthetic

Wool right off the sheep is extremely dirty and has to be washed and then "picked" or pulled apart so the dirt and debris fall out and the fibres are loosened. This large basket of clean wool is picked and ready to be carded.

traditions, distinctive work patterns, and their living and working spaces. The fabrics they made also had identifiable characteristics, both in construction and design.

The story of this vital industry and its growth, from the work of home-based men and women, who spun to supply their own families — and perhaps to earn some money on the side — to the factories that took over these time-consuming chores, provides fascinating detail into the lives of Canadian producers and consumers as cloth-making moved from the farm to the factory.

The addition of drive belts and wheels to a spindle turned on its side sped up the spinning process; the spinner walked back to spin the yarn and forward to wind it on the spindle. When the spindle was full, she used a reel that wound the yarn into a skein and measured it.

1 Fashioning Homespun

While the Vikings may have been the first Europeans who came to Canada, they were only the beginning of a long trend that resumed early in 1605, when the French established their first successful settlement. Port Royal Habitation was situated in a protected bay discovered by Samuel de Champlain on the north shore of what is now the Annapolis Basin, opening off the Bay of Fundy in Nova Scotia. Some of the

Different breeds of sheep came from various parts of Europe. Here, a group of sheep huddle together at Black Creek Pioneer Village.

Top: Picking the washed wool separated the fibres to make carding easier. These jobs could be done by children.
Above: There was always time to play with a new lamb.

first communities in North America were housed in forts and garrisons like Port Royal, where the population, largely male and transient, would have worn and used clothing and textiles made in Europe. Not until family groups arrived and began farming was there any significant domestic cloth production.

Over the course of the seventeenth century, the countryside around Port Royal became a prosperous agricultural community. Populated largely by French immigrants, the region became known as Acadia, and its residents as Acadians. Not surprisingly, the Acadians brought many of their traditions with them from France, including textile-manufacturing skills and clothing styles. Once they had successfully built dykes to reclaim the marshland on which they settled, the Acadians began farming in earnest, and soon they were producing most of the food and goods they needed to live well. As their numbers grew, they even generated an agricultural surplus, which they exported in exchange for European goods — including imported cloth. However, they made most of their textiles from flax they grew themselves and from the wool of sheep they raised. In 1717, a French official at the fortress of Louisbourg commented: "These French Acadians are hardworking by nature ... they are born smiths, joiners, coopers, carpenters, and builders. They themselves make the cloth and the fabrics in which they are dressed." The community was not isolated from the rest of the world and traded with Europe and New England for manufactured goods and dyestuffs like vermilion, which they combined with materials of their own production.

In cold northern climates, sheep's wool is essential for warmth; and the animals are also a source of food. Although evidence from this early period is scarce, it is likely that most Acadian households had a few sheep. The men would shear them annually, and their wives and daughters would process the fleece to knit or weave into cloth. Once the wool was cleaned, the fibres had to be prepared for spinning with tools called "cards." Carding was usually the work of women and children, who brushed the wool until there were no more

Top: Acadian spinning wheels were unique because the flyer mechanism protruded outside the maidens rather than sitting between them.
Above: Skeins of wool

Knitting was a portable craft that could be done while standing or walking, sitting in the kitchen to wait for food to cook or at a sociable occasion while visiting friends.

tangles and all the fibres lay parallel, then they rolled it off the card ready to spin. This is a very time-consuming step in yarn production, and the tools and technology did not change until the early nineteenth century, when water-powered carding mills took over the work. Once a spinner had accumulated enough rolls she could begin spinning.

As mentioned, a yarn can be made with a simple spindle — a shaft with a weight that is rotated between the thumb and forefinger — or with a spinning wheel, which speeds up the process by turning the spindle on its side and adding a drive wheel and belt. Wheels like this were suitable for spinning short fibres like

wool or cotton (flax fibre was too long) and were called "great wheels" or "walking wheels," as they were large and the spinner had to walk back and forth to make a yarn. Another improvement was the addition of a bobbin-and-flyer mechanism set between two upright posts ("maidens") that simultaneously spun a yarn and wound it on the bobbin. The entire unit was called the "mother-of-all." These smaller wheels took up less space because they were smaller and a foot treadle turned the wheel so the spinner could work sitting down. Depending on the diameter of the wheel one could spin wool or flax on devices like this. (Large wheels generate more

Above: When the flax was ripe, the entire stalk, root and all, had to be pulled out of the ground to obtain the longest fibre possible.
Right: It was back-breaking work for those who pulled the flax from the ground.

twist and are suitable for short fibres, while long fibres like flax need a smaller wheel ratio to twist it less.) Some Acadian spinning wheels had a unique characteristic: they set the flyer and bobbin outside the uprights, or "maidens," rather than between them.

In addition to wool, flax was used extensively at this time. It required a lot more effort than wool to turn it into yarn, but the large Acadian families (averaging seven to ten children each) made the work easier. Many people would have grown small amounts of flax for their own use in household linens and clothing — as little as

Men usually did the physically demanding work of breaking the flax, using a tool called a "flax break" to loosen the outer bark and make the fibre accessible.

one quarter of an acre could provide much of a family's needs. Planted in the spring, and harvested midsummer, flax took almost a year of arduous, if intermittent, preparation before it was ready for the loom. First it had to be dried for two weeks, then the seeds at the top of the

Women often spun flax outdoors.

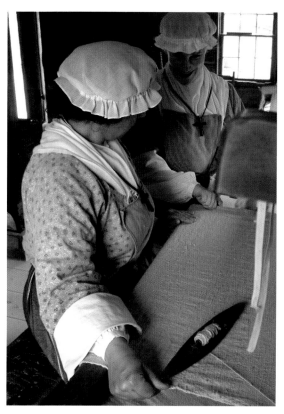

The simple looms the Acadians used were more than adequate for the production of household linens like the piece in this loom. It may have been destined to become a bedsheet.

plant were removed by pulling them through a coarse comb (a process called "rippling"), after which the outer bark was exposed to moisture (dew or water) to rot it for two weeks to a month, and then dried again. The bark was further weakened by breaking it with a tool called a "flax break," and removed with a "skutching knife." The final step, called "hackling," involved pulling the strands of flax through a series of comb-like devices called "hackles," which straightened the fibre so it could be draped, like long flaxen hair, and secured on a

"distaff" to keep it from tangling during spinning. All this work would have occupied many hours throughout the year for hard-working Acadians, but when they were done they would have had a long, strong fibre to spin with a small wheel perhaps equipped with a distaff.

Whether they used wool or flax, or combined the two in a single piece of cloth, Acadians used a simple wooden loom capable of weaving only the most basic fabric.

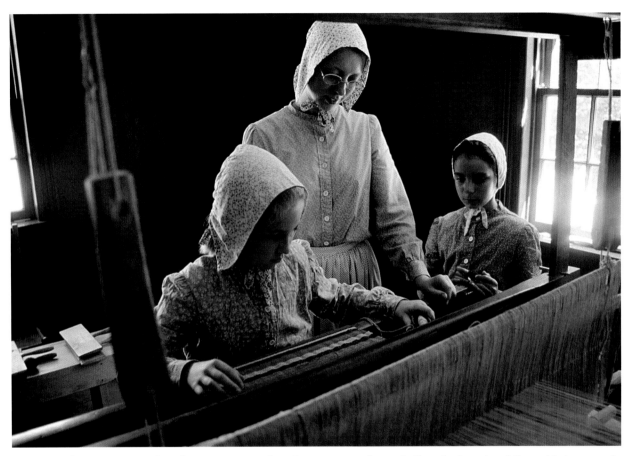

This weaver is teaching her two young students how to weave the vertically striped wool and linen skirting worn by Acadian women.

Nevertheless, they produced attractive, durable fabrics. Because handmade cloth is so time-consuming to produce, it was used until it wore out or was recycled. As a result none has survived. Nevertheless, descriptions in historical records give us a sense of some of the fabric produced in this early period, especially women's costumes. Their skirts followed the French tradition of vertical stripes, perhaps in combinations of red, white, and blue, or of black, blue, yellow, and white, and their blouses were made of linen. For special occasions they may have combined their homemade clothes with some of imported fabric. Men wore simple dark woollen trousers, and they too would have had shirts made of linen. On their feet they wore wooden shoes or moccasins.

Some of what we know about the textile traditions of the Acadians during this early era comes from our knowledge of how they lived

Everyday clothing for Acadians consisted of linen and wool cloth of their own manufacture, caps and socks that they knit themselves, and wooden shoes.

later. To simplify a very complicated story, until 1755, the Acadians prospered and coexisted with the changing English and French regimes, but when the British demanded an end to their neutrality in 1754, they refused to take the required oath of allegiance to the British Crown. As a result, the British removed over twelve thousand people from their Acadian homeland, in the process destroying their farms — an action now known as the Great

Once wool cloth was removed from the loom, it had to be shrunk or felted in a process called "fulling." The French tradition involved immersing the wet, soapy fabric in a trough, where men used mallets to beat it while they kept it moving until it was the desired thickness.

Expulsion. When the political situation between the French and English resolved later in the eighteenth century, the Acadians were allowed to return to Canada, and they settled in areas of Nova Scotia, Prince Edward Island, and New Brunswick, where they continued to farm and make their own cloth well into the twentieth century. Unlike those of the first period of settlement, textiles from the nineteenth century did survive and probably resemble what was made earlier: striped wool and linen skirt lengths; bedding made from recycled material, attractively patterned in bands of brown, red, white, and blue; and elegant all-white bed coverings known as *couvertures de mariage*, or "marriage coverlets," all woven on simple looms.

The French did not restrict their early settlements to Nova Scotia but moved further up the St. Lawrence as far as what is now Quebec City and Montreal. For most of the seventeenth century, the St. Lawrence region was settled mostly

Even when early Canadians made most of their own textiles, they could always buy imported fabrics, trimmings, and threads.

by men engaged in the lucrative fur trade or in mission activities. By the beginning of the eighteenth century, however, family groups in small numbers gradually moved beyond the trading posts and began to farm. The climate of the St. Lawrence, with its short hot summers and long cold winters, made agriculture difficult. Nevertheless, the *habitants* were able to eke out a living from the land.

By the middle of the eighteenth century, the population had grown sufficiently that a family could largely subsist on the products of its own labours, including clothing and textiles. Much like the Acadians, French-Canadian women, with the help of their children, spun wool and flax and wove it into cloth. Given the frigid winters, wool was especially important. To maximize its insulating qualities this fibre needs to be shrunk and brushed to make it denser and fluffier, and hence more able to trap air. This final step was men's work, as it involved pounding and moving yards of wet,

Most early Canadians, whether they lived in French or British regions, in the town or in the country, had access to imported clothes, linens and furnishings.

soapy woollen cloth in a trough carved out of a hollow tree trunk, using long-handled wooden mallets. Fabrics shrunk this way made long-lasting and warm clothing and blankets.

Although textiles from this era have not survived, documents have. Of particular use are the formal agreements that ensured members of the older generation would be looked after if they followed the French-Canadian tradition of transferring their farms to their children before they died. For example, in 1791, Joseph Blanchard and his wife, Marie Daigle, both of St-Ours, gave their holdings to their son, who agreed among other things to provide them with a furnished room, do their washing and mending, and supply them with three pounds of wool, four shirts of homespun, a complete suit of work clothes, sheets and shoes as needed,

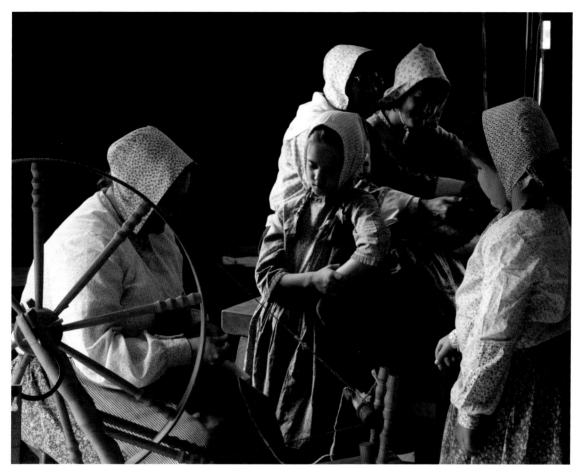

In French and Acadian households, textile production was women's work. Young girls learned how to spin from their older sisters, their mothers or their grandmothers.

and a set of Sunday clothes purchased from the merchant every three years. This list of necessities, with its reference to store-bought clothing, reminds us that, while these families may have manufactured a lot of cloth, they were not completely removed from a wider market.

Textile production in French Canada was almost exclusively rural and the province of women. Moreover, it continued, and even expanded, throughout the nineteenth century, though many more households owned spinning wheels than looms — not surprising, since it takes approximately six spinners to provide enough yarn for one weaver, and since knitting was a widespread activity. An 1830 report by Surveyor General of Lower Canada Joseph Bouchette to the British king gives some measure of the extent of domestic cloth production. He claims there were over thirteen thousand looms in Lower Canada, producing substantial amounts of country cloth (*étoffe du pays*), flannel and homespun (*petite étoffe*), and linen. Bouchette says that, while there was some production for market, most people were producing for themselves alone, though by this time *habitants* were increasingly purchasing cheap English goods. His description of the dress of rural French Canadians is evocative:

> *The gray* capote *of the* habitant *is the characteristic costume of the country. This* capot *is a large coat reaching to the knee, and is bound round the waist by a sash, which sash is usually the gayest part of the Canadian's dress, exhibiting usually every possible bright colour within the power of the dyer. This, with a straw-hat in summer, a* bonnet rouge *or a fur cap in winter, and a pair of mocasins made out of sole leather, complete the dress of the peasant. The women are clothed nearly after the fashion of a French peasant: a cap in place of a bonnet, with a dark cloth or stuff petticoat, a jacket (*mantelet*) sometimes of a different colour, and mocasins, the same as those of the men, form their every-day dress. On the Sunday they are gaily attired, chiefly after the English fashion, with only this difference — where the English wears one the Canadian girl wears half a dozen colours.*

All the fabrics discussed above were woven on the simplest loom, capable of making only the most basic weave structure, called plain weave or tabby.

As the nineteenth century progressed, water-powered carding, spinning, and weaving mills began to dot the countryside, but domestic spinning and weaving in Acadia and French Canada persisted longer than it did in other parts of Eastern Canada.

2 Traditions and Transition

The end of the eighteenth century and the beginning of the nineteenth was a turbulent time in the British Atlantic world. It was an age of revolution — agricultural, political, and industrial — which generated a flow of immigrants to Canada from Scotland, England, Ireland, Germany, and the newly formed United States. They brought their textile traditions with them, some of which survived intact, while others had to be

Regardless of ethnic background — French, English, Scottish, or German — children were an important source of labour when textiles were made at home. These girls are picking wool and carding it at McDermid House at Upper Canada Village.

Left: Fuller's teasels were burr-like plants used instead of the imported bent wire teeth on hand cards and were also used to brush woven wool cloth to raise a nap. Above: Wool ready to be cleaned.

adapted to suit the new environment.

Prince Edward Island, Nova Scotia, and Cape Breton Island were the first destinations for many Highland Scots; and later, Lowlanders joined the migration and settled in parts of what are now Quebec and Ontario. The first Germans went to Lunenburg, Nova Scotia. The English, many of whom came from the British colonies south of the border after the American Revolution of 1776, settled all over Nova Scotia, New Brunswick, and Ontario. Another group of Loyalists were the Mennonite Germans from Pennsylvania, who went primarily to southwestern Ontario.

Although in the eighteenth century cloth production was at different stages of industrial advancement in Europe and America, usually spinning was done by women and men did the weaving. But in Canada, a country more heavily populated by trees than by people at first, settlers of all nationalities had to go back to the basics, regardless of how developed textile manufacture may have been in their homelands.

Choosing to leave one's country of birth was a wrenching decision, though the Scots, for example, found a more welcoming landscape,

In the new carding mills, huge quantities of mechanically cleaned and picked fleece could be loaded into water-powered carding machines.

with lots of trees, water, fish, and a longer growing season than the one they left. Log cabins supplanted the stone, sod-roofed houses they had left behind and, like the Acadians, they lived off the land and made much of their own cloth. Others, such as the Americans, had abandoned established farms or city homes and moved to the Canadian frontier, where their first houses were also built of logs.

Families had to think carefully about what they brought with them, and often the spinning wheel was among their cherished possessions. A nineteenth-century novel about a young Canadian girl describes one such wheel:

It had been part of the marriage outfit of Shenac's grandmother before she left her Highland home. It had been in almost

Families who brought wheels with them could begin to make yarn for knitting warm garments fairly quickly after their arrival. Spinning wheel making was a highly skilled craft. It took a while for a community to be able to support a local cabinetmaker who had the skills to make new wheels.

Sheep usually have their fleece sheared off only once a year. As it grows, it accumulates mud, straw, grass, and dung, which must be removed before the fibre can be processed for spinning a fine, consistent yarn.

constant use all these years, and bade fair to be as good as ever for as many years to come. There was no wearing it out, or putting it out of order.... It was a small, low, complicated affair, at which the spinner sat, using both foot and hand. It needed skill and patience to use it well.

A wheel like this could only be made by an artisan practised in the craft of wood-turning and joining, and it would have been expensive. In early frontier settlements, these craftsmen were not easily available, and neither was cash. Women who brought their wheels with them could quickly begin making yarn to knit the mittens, stockings, "Guernsey frocks" (sweaters), comforters, and nightcaps needed during the long, cold Canadian winters.

Cloth-making in early nineteenth-century Canada wasn't very different than it had been in the preceding century, and no matter where people came from, they shared the same procedures for fibre preparation. Flax and wool remained the raw materials, though technology, in the

Once washed, the fleece had to be picked to remove solid debris and any dirt that remained. Formerly the work of children, this dirty, unpleasant job was taken over by water-driven machines like this after the advent of carding mills.

form of carding mills, was about to transform the centuries-old process. When they first arrived, the Scots in Cape Breton used small Highland sheep that shed their wool, but by the mid-nineteenth century the introduction of new breeds meant that their sheep, like oth-

ers, had to be sheared. Canniff Haight reminiscing about his life as a young man in rural Canada in 1850 recalled the work involved in wool preparation:

In June came sheep-washing. The sheep were

A pair of hand cards showing the bent wire teeth used to brush the wool fibre. It took many women and children a great deal of time to card enough wool for a spinner to make sufficient yarn for weaving.

driven to the bay shore and secured in a pen, whence they were taken one by one into the bay, and their fleece well washed, after which they were let go. In a few days they were brought to the barn and sheared. The wool was then sorted; some of it being retained to be carded by hand, the rest sent to the mill to be turned into rolls.

English settler and writer Catherine Parr Traill elaborated on this stage of fibre prepara-tion in her guide for emigrant housekeepers:

All dirty wool is thrown aside, and those who are very careful will sort the coarse from the fine in separate parcels. The wool when picked is then greased with lard, oil or refuse butter, which is first melted then poured over the wool, and rubbed and stirred with the hands until it is all greased: about three pounds of grease is allowed to seven or eight pounds of wool, it is then fit for the carding mill.

When the spindle or bobbin of the spinning wheel was full, the yarn had to be wound off and measured. The simplest device for this was a niddy noddy, as seen above left. On the right is an umbrella swift that held a skein of yarn to be wound onto warping spools or into a ball.

By the 1820s, carding mills were springing up everywhere, ultimately eliminating the long, social winter evenings when family and friends would get together to card the wool. Even though the nearest mill might be miles away, and the service paid for, it was still more cost-effective than carding by hand, and workers freed from this task could attend to the many other jobs on the farm.

Carding is a crucial step in wool spinning, as any dirt or lumps left at this stage result in uneven yarn. Water-powered carding machines, consisting of a series of cylinders covered with teeth of bent wire, produced a long, thin roll or "roving," superior to the short thick rolls fashioned by hand. Mill-spun roving is especially good for spinning on the walking wheel. A contemporary observer describes what happened once the roving was returned from the mill, at which time he says,

> *the hum of the spinning wheel was heard day after day, for weeks, and the steady beat of the girls' feet on the floor, as they walked forward and backward drawing out and twisting the thread, and then letting it run upon the spindle. Of course the quality of the cloth depended on the fineness and*

evenness of the thread; and a great deal of pains was taken to turn out good work.

Catherine Parr Traill recalled it as a very becoming activity for young women, because the constant walking showed off their figures. Working continuously, a proficient spinner could produce about a pound of wool a day, but it took six to eight pounds to weave a blanket.

Once the spindle was full, the yarn had to be wound off into a skein with a reel that could be as simple as a hand-held "niddy noddy" — a wooden device consisting of a shaft with two crosspieces on the end. Its name derived from the nodding motion it made while in use, and generated a descriptive rhyme: "Niddy noddy, niddy, noddy; two heads and one body." A more sophisticated reel, called a "clock" or "click" reel, contained a gear-and-peg mechanism to measure the length of each skein, crucial knowledge for weaving. A pointer that looked like the face of a clock indicated the number of revolutions, and the peg clicked to indicate a specific yardage.

Spinning wheels such as this "Irish Castle Wheel" saved space in small homes.

Wool right off the spinning wheel consisted of a single strand suitable for weaving (singles), but, if it was for knitting, it had to be plied together with a second strand and spun in the opposite direction to create a yarn that did not twist. This ensured that the finished garment would hang straight. Once in skeins, the yarn could be dyed. In the early nineteenth century, there were professional dyers. In 1827, for example, the Nova Scotia General Assembly gave Gideon Harrington £241 "to enable him to extend his Manufactories for carding, fulling, and dyeing, ... by adding thereto a cold Indigo dye Vat for dyeing cloth a permanent blue." But many spinners who made their own yarn coloured it at home, using both local and imported dyestuffs. Carding the wool of black and white sheep together created "sheep's grey," used for men's homespun work clothes and socks. The bark, roots, nuts, and leaves of plants delivered an array of colours. Butternuts and walnuts yielded a deep, dark brown; goldenrod flowers created a nice yellow and, when combined with

Black walnuts were a local dye source that produced a dark colourfast brown. The iron pots that these women are using would have intensified the colour as the skeins of spun yarn were left in the warm dye bath.

imported indigo, a bright green; logwood, also imported, produced a permanent black; onion skins made a gold or light brown; madder root, another imported dyestuff, gave red. Imported dyes could be obtained from the general store or perhaps from a peddler passing through, while local colourants came from the garden or the woods.

Dyeing was a complicated affair, and required chemical knowledge. It was not as easy as simply soaking the dyestuff and the fibre in water. A mordant such as alum (aluminum potassium sulfate) was needed to set the dye. Moreover, the colour obtained from a single dyestuff could be varied by incorporating different chemical mordants into the dye bath. Iron, for example, darkened colours, so it was important to know that heating the wool in an iron pot would create a different shade than a copper pot. Other chemicals had to be pur-

When the dyeing process was completed, the wet skeins of yarn had to be dried.

chased. Blue vitriol, verdigris, copperas, or cream of tartar could enhance and change colours, but the dyer required knowledge of complicated dye recipes in order not to ruin the yarn that had taken so long to make. Wool was easier to dye in an array of colours than was cotton or linen. Indigo produced a colour-fast blue on linen, otherwise linen tended to be left natural or be bleached white. It is important to remember that despite all the work involved in fibre preparation, spinning, and dyeing, it represented only a fraction of what women had to do, and had to be interspersed with food preparation, childcare, cleaning, laundry, and gardening, with none of the labour-saving appliances we enjoy today. No wonder some women preferred to pass the dyeing on to the local weaver or dyer!

Once dyed, the yarn was ready to be knitted or woven into cloth. Catherine Parr Traill admonished young girls, regardless of class, to learn to knit if they did not already know how. Women who knit, says Traill, not only supplied the needs of their family, they could also sell their surplus goods:

Men's socks sell at one shilling and six pence to two shillings and three pence, according to their goodness … the second or even third rate wool, knitted up, can be made more profitable than the best wool sold in the fleece, and children and women will earn many a dollar if they are industrious, in the evening, between twilight and candlelight.

Wool could be dyed in the fleece, before it was spun, or in the skein, post spinning. Here is a basket of two-ply yarn to be used for knitting, dyed in indigo blue, cochineal red, and undyed sheep's grey.

Traill further remarks on the sociability of the work, as young women took their knitting out with them when they visited friends. According to her, in addition to warm clothing "many persons knit cradle-quilts, and large coverlets for beds, of coloured yarns, and among the town-bred young ladies, curtains, tidies for sofas, and toilet covers, of all sorts and patterns, are manufactured with the knitting-needles, and cottons of suitable qualities." Yarn not used for knitting was woven into cloth.

In Europe and parts of the United States and Canada, early in the nineteenth century, weaving was often a commercial venture done by skilled male artisans. But in the new Canadian frontier settlements the men had to focus their energies on the heavy toil of clearing land and establishing farms, so frequently textile making was taken over entirely by women (though some specialized, male weavers operated small businesses). With little time for a traditional apprenticeship many women of Scottish and English extraction

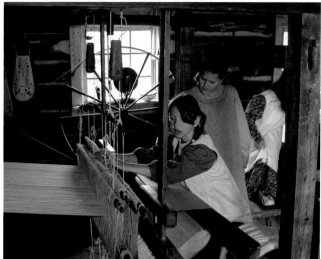

wove only the simplest fabrics, though their looms were slightly more complex than those of the Acadians and French-Canadians. With four shafts instead of two, they could make cloth with more complicated weave structures and designs.

Creating even basic cloth required many more tools and a lot of work before the weaver could sit down at the loom. To make the warp she had to know the length of the finished fabric and its final use in order to decide on its density (determined by the number of warp ends per inch). For example, a warm woollen blanket might have twenty ends per inch, homespun for a man's work clothes thirty-two, and a fine linen tablecloth might have forty. She put the skeins of yarn onto a "swift," a holder that allowed them to be unwound onto spools, the number of which was controlled by the required warp ends per inch. When enough spools were filled, they were

Top left: A skein of handspun linen and a shuttle
Top right: Young girls who were learning to weave would be allowed to try their hand at weaving plain, simple fabrics first.
Above: Loom in operation

Here the weaver pushes back the beater that holds the reed with her right hand while she inserts the shuttle holding the weft yarn into the space between the separated warp threads known as the "shed."

secured to a rack so the weaver could unwind the threads smoothly, under even tension, around a series of pegs set at measured distances apart on a warping board. If the tension was not uniform, the finished cloth would sag at odd places. She then tied the finished warp together at intervals before removing it in a chain that would guarantee that the hundreds of threads did not tangle. It took two people to wind the warp on the loom's warp beam, again making sure the tension was consistent. A series of shafts in the centre of the loom

The mid-nineteenth-century invention of the sewing machine saved some of the work of hand stitching.

contained heddles — loops of yarn with a smaller loop or "eye" in the centre. Each warp was threaded through a heddle eye in a pre-determined sequence to create the desired design. Once threaded, the warp was tied onto a cloth beam and the weaver was almost ready to begin, but first she had to prepare her shuttles to carry the weft yarn across the warp.

Another wheel — a quill wheel or bobbin winder — wound the weft thread onto the shuttle's bobbin, a job that might be done by the children in the family. A simple cloth of one colour needed only one shuttle going back and forth, as alternate warp threads were raised and lowered through the action of the weaver's feet on treadles. A plaid or a check, however, would require two or more shuttles, each holding a different colour. The more colourful and complicated the weave pattern, the more the weaver had to concentrate, preferably working in uninterrupted blocks of time, something women with small children

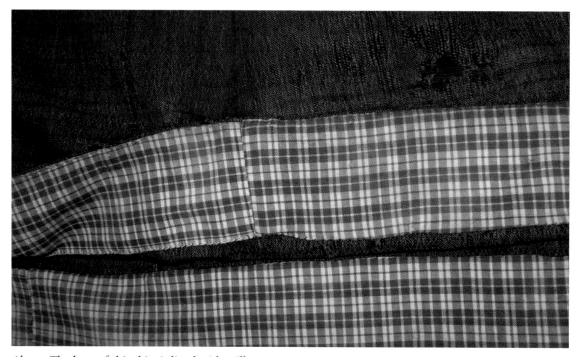

Above: The hem of this skirt is lined with mill-woven cotton that would have allowed it to be lengthened. Right: Homespun clothing was durable but scratchy. This man's work shirt is lined with soft machine-made cotton, and the collar, too, is made of cotton to reduce chafing around the neck.

may have found difficult. Lack of training then, and the pressures of child care and other housework, meant that many women wove only relatively simple homespun fabric.

Recording her observations of life in New Brunswick in 1845, after living there for seven years, Emily Beaven described the kinds of things one housewife wove:

> *Her dress is home-spun, of her own manu-*
> *facture, carded and spun by her own hands,*
> *coloured with dyestuffs gathered in the*

Admiring a handsome handspun, dyed and knitted red petticoat at Upper Canada Village

Socks, mitts and warm underwear were among the staple knit goods produced from handspun yarn.

woods, woven into a pretty plaid, and neatly made by herself. This is also the clothing of her husband and children; a bright gingham handkerchief is folded inside her dress, and her rich dark hair is smoothly braided.

Children's clothes were not much different than those of their parents; indeed, adult clothes were often cut down for the smaller members of the family. But what did they wear underneath? It is difficult to know for sure, because people did not discuss such things at the time, but a few surviving examples of women's drawers and petticoats suggest that they, too, were of rough homespun, sometimes embellished with red wool embroidery (the friction of the rough wool against the skin would generate warmth on a cold trip to the outhouse). The drawers were constructed with a back that could be dropped independently or with an opening that went from front to back, making them practical garments in homes that lacked plumbing and insulation. In addition to clothing that might be all wool, or a

The handwoven overshot coverlet folded at the end of the bed in an early twentieth-century room at Highland Village, Cape Breton, was used in conjunction with an all-white cotton bedcoveringI that was likely imported from England.

combination of wool and linen or cotton, called linsey-woolsey, women wove wool blankets, and checked winter sheets, perhaps some plain linen towels and sheets, and, if they had time and enough skill, a special overshot coverlet. Families without a loom, or those who wanted special items they couldn't weave themselves, could, money permitting, commission the services of more specialized weavers.

3 The Commercial Weavers

In addition to the women who made textiles for their families, in most regions of eastern Canada, in the first half of the nineteenth century, there were professional or commercial weavers (mostly, though not exclusively, men). While they, too, wove basics like homespun and simple functional cloth according to their customers' needs, they also had the training and skills to make more elaborate fabric goods, many with distinctive designs reflective of European traditions. Some of these artisans even developed a specialty trade in items like coverlets. By the middle of the nineteenth century, the number of commercial weavers increased along with Canada's developing economy and accelerated immigration. At the same time, changing technology continued to penetrate the countryside, as water-driven spinning and weaving mills extended the impact of the carding mills, taking over more of the work formerly done by hand. During this transitional period, household and commercial hand weaving co-existed with a small, but growing, factory production. At the same time, rural residents

Imported cotton wound on spools. It was used to make warps.

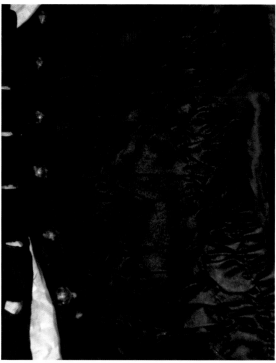

Left: Some villages had a dressmaker who could sew the expensive imported fabrics used in clothing for the wealthy or for special occasions.
Above: Detail of a dress made of imported silk fabric by a dressmaker in Peterborough, Ontario, about 1850.

bought imported textiles, mostly of cotton, to supplement locally made materials.

Even the most remote backwoods communities in the first half of the nineteenth century, where homespun was a necessity,

were not completely isolated from the retail world or from ideas of fashion. Whether their needs were supplied by a travelling peddler, a local storekeeper, or a trip to a village or town, rural residents could buy an array of textile-related goods: notions, such as needles, thread, ribbon, lace, and bindings; cottons like calico, muslin, and gingham; fine linens like cambric; and even luxury fabrics such as silk and velvet. English cotton cloth, much of it printed in a variety of colours, however, was the fabric most people bought. A woman might have tucked a blue-and-white-checked gingham neckerchief into her bodice for comfort and modesty, or lined the collar of her husband's homespun shirt or the waistband of his trousers with a small piece of imported cotton to prevent chafing. As the nineteenth century progressed, and frontier communities became more established, more and more garments, including dresses, aprons, shirts, and trousers, were made completely of cotton — especially in the hot summer. This versatile fabric was attractive and could be easily washed. At the very least, most people owned a cotton handkerchief.

Surprising as it may seem, imported cloth was often cheaper than locally made material. An 1817 statistical account for Upper Canada observed that, in some regions, "it was full as cheap to go to the store and buy English broad cloth as to make homespun, for this obvious reason, that by the time it went through the hands of the carder, the spinner, the weaver, the fuller, and the dyer, it cost more per yard than the English, and generally of inferior quality." While some of the processes could be done at home, many steps like carding, fulling, and dyeing, and increasingly weaving and spinning, were done at a mill or by a commercial weaver and cost money. Luxury fabrics such as silks and velvets were not made in Canada, however, and a few people, usually the wealthiest in a community, bought these from local storekeepers, perhaps for a special occasion like a wedding. But it wasn't just fancy imported cloth that

Specialized band looms were used to weave narrow tapes for things like apron strings, garters, and ties for grain sacks or mattress covers.

some rural Canadians coveted; one very special, locally made item — the bed coverlet — took pride of place in many country households.

The prized coverlets were both beautiful and functional, but their manufacture required specialized skills and equipment that were beyond the capacity of early power looms, and imports were too expensive. Emily Beaven's 1845 account of her visit to a backwoods home in New Brunswick could apply to almost any of the Scottish, English, or German frontier home-

A professional weaver setting up his loom, equipped with a flying shuttle, to begin a new project. Behind him are samples of the coverlets he wove, from which his customers could select the pattern of their choice.

steads in eastern Canada. Her description of the one-room house in which many young families began their lives helps to explain the importance of these attractive bed coverings. Beaven notes that she visited a house in which the single chamber, in addition to serving as parlour, kitchen, and hall, was

> the state bedroom as well, and on the large airy-looking couch [bed] is displayed a splendid coverlet of home-spun wool, manu-

factured in a peculiar style, the possessing of which is the first ambition of a backwoods matron, and for which she will manoeuvre as much as a city lady would for some bijou of a chiffionier, or centre table.

Beaven further observed that the owner had

> gained hers by saving each year a portion of the wool, until she had enough to accomplish this sure mark of industry, and of

The shuttles were tipped with steel so they did not wear out as they shot with force and sped back and forth from one box to another.

getting along in the world; for if they are not getting along or improving in circumstances their farms will not raise sheep enough to yield the wool, and if they are not industrious the yarn will not be spun for this much prized coverlet, which, despite the local importance attached to it, is a useful, handsome and valuable article in itself.

These cherished coverlets provided colour and warmth for beds that were often on public view in one-room cabins. Even when homes became larger and the bedrooms separate, they continued to be treasured.

Woven in various patterns and constructions, coverlets were among the goods that country customers ordered from skilled market weavers or professional weavers. Communities like those in rural New Brunswick or Cape Breton Island, where many women made their own simple homespun fabric, usually had a more accomplished female weaver one could

Professional weavers often used large warping mills like this to make long warps for several projects. The warp yarn is wound on spools that are placed on a rack, then threaded through a device to keep the tension even while winding.

pay to make a special coverlet. In addition to market weavers like these women, some professionally trained Scottish, American, and German cloth workers either set up as full-time craftsmen or combined part-time weaving with farming. These men, many of whom settled in Ontario, usually had more than one loom housed in a separate workshop and wove on a custom-order basis. They often used yarn provided by their customers, though some made

yardage for sale in local stores alongside the imported fabrics. Unlike the women whose customer base was established through word of mouth and local knowledge, the clientele of the professional weavers came from further away. To build their businesses, these men might promote their skill in local newspapers or distribute printed leaflets advertising their products. For example, when he set up his business in 1836 in southwestern Ontario, Mennonite

The foot pedals of the loom were called "treadles." Weavers had to depress them in a specified sequence to achieve the desired pattern.

weaver Samuel Fry's handbills announced that he "respectfully informs the Inhabitants of the Niagara district, that he is prepared to WEAVE all kinds of PLAIN and FANCY **COVERLETS**, DIAPERS &c. at reduced prices, in a workman-like manner, and on reasonable terms." Although prepared to manufacture a variety of fabrics, clearly Fry, who had just returned from Pennsylvania where he learned his craft, expected his "fancy coverlets" to be a big component of his business.

Many coverlets, regardless of design, were made of white factory-spun cotton imported from New England and combined with locally spun and dyed wool yarn (most often dark blue or red). The cotton formed the foundation fabric and the wool created the pattern. The various types of coverlets had names like "Overshot," "Summer and Winter," "Twill Diaper," "Star and Diamond," and "Bird's-eye" that suggest how they were constructed or how they looked. In "overshot" coverlets, for example, the wool weft shot over multiple cotton warps in a predetermined sequence to form a

Commercial weavers usually had a workshop and more than one loom.

pattern. Floating on top of the cotton foundation, the wool yarn created small air pockets that provided insulation and made these attractive bed coverings extremely warm for cold winter nights. Weavers working on looms with only four shafts could create a surprising number of different designs by varying the sequence in which they threaded the warp ends through the heddles. It is likely that the Canadian overshot coverlets originated in Scotland, though this weave structure can be found elsewhere as well. A great many were made in eastern Canada, and these distinctive textiles have become almost synonymous with our pioneer past. Some of the overshot patterns had standardized names that were also suggestive of their appearance, like "Chariot Wheel," "Snail's Trail," "Turkey Tracks," or "Nine Roses." Even history played a part, as seen in a Cape Breton Island example called "Wellington's Army in the Field of Battle."

While overshot coverlets may have been especially popular with people of Scottish origin, another type called "summer and winter" was among the English traditions brought to Canada by United Empire Loyalists. They were

To save time and minimize the possibility of making a mistake when starting a new project, the weaver left the threads from a finished piece in the heddles and tied the new warp onto the old one, then pulled it through.

woven on looms that required six shafts, and the weft floats were shorter, with the result that these fully reversible coverlets were darker on one side (used in winter) than the other (used in summer), a practical function in a society where fabric was hard to come by. Generally any textile that calls for more than a four-shaft loom would have been made by weavers with a lot of skill and training. Numerous coverlets found in German-speaking settlements in Ontario, for instance, were made on very complex looms, some with as many as sixteen shafts, the simplest being "Bird's-Eye"; "Star and Diamond" and "Twill Diaper" (a damask-type construction) were much more complex. Many beautiful examples woven by Samuel Fry have survived, almost always with a red, white, and blue palette applied in varying combinations. Not only were these textiles used on beds, they may have later been woven as horse blankets, especially when the market for coverlets diminished. Today it is difficult to understand the value of a horse to nineteenth-century

Jacquard looms used a series of punched cards and hooks to make the design. In recognition that these were the very first computers, this loom is still in operation at the Ontario Science Centre.

families for both work and transportation. Horses were also on public display as they stood hitched to a wagon or carriage waiting for a family while they went visiting, shopping, or to church. What better way to communicate status and the value of these animals than to cover them in specially ordered blankets? Many horse blankets made in a twill diaper by professional weavers who emigrated from Germany were so treasured over the generations that they are now in museums. English Loyalists, too, wove bed coverlets in this multi-shaft weave structure. The term "twill" designates how the fabric was woven, and the term "diaper" indicates the design. While today we think of the latter term only as it relates to babies' diapers, originally it specified a textile with a small, repeating overall design. Woven with several colours of wool and cotton, it made a beautiful bed or horse covering. Woven in linen or cotton, the diaper construction added extra absorbency to

A jacquard coverlet in the process of being woven. The mechanism allowed individual threads to be manipulated, thereby creating very realistic designs, unlike those produced on regular looms, which used a series of geometric blocks to create the illusion of curves.

the fabric, making it practical for towels or baby garments.

A few very specialized weavers, who were trained to weave carpets in Scotland, or fine linens like damask in Ireland and Germany, continued to weave after they arrived in Canada, but they had to adapt their skills to a very different market. Some of these men had been trained to work the new jacquard loom that was perfected in France in 1801 for weaving silks with lifelike designs. These looms were, in effect, the first computers, as they had an attachment that employed punched cards interacting with a series of hooks to make the pattern. Because individual threads could be manipulated, it was possible for the first time

Overshot coverlets were prized possessions in early farm households.

to make curved, hence more realistic, representations of flowers, birds, animals, and buildings (Even using multi-shaft looms, all designs were previously constructed in a series of geometric blocks that gave only the illusion of curves.) With little demand and inadequate raw material to weave fancy carpets and fine linens in Canada, however, these emigrant weavers adapted their skills to the coverlet market.

Despite their mechanical complexity, the wooden structure of jacquard looms differed little from the shaft looms used for simpler fabrics. Almost all Canadian hand looms, whether used by professional or home weavers, were narrow, so most blankets, sheets, and coverlets had to be woven in two widths and seamed up the middle; a weaver's expertise was apparent in how well the patterns matched. There were several reasons for the narrow looms. One was space, though the jacquard head added a lot of height to the loom — indeed, one weaver, John Campbell of Komoka, Ontario, had a workshop with an especially tall ceiling to accommodate his. But the main determination of fabric width was the weaver's arm span as he or she threw and caught the shuttle. Most fabric handwoven in Canada was about twenty-seven to forty-five inches in width, unless made on a broad loom that required two people to throw the shuttle back and forth to each other (not practical in a society where labour was scarce) or a mechanized fly shuttle that also sped up the weaving. Many professional weavers owned at

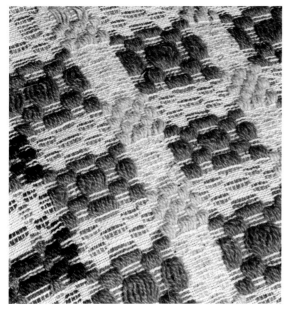

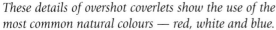

These details of overshot coverlets show the use of the most common natural colours — red, white and blue.

least two looms, one for complex fabric, the other for simpler cloth, on which they produced custom orders with cotton imported from New England and wool yarn provided by their customers.

Young Canadian girls expended a great deal of effort to accumulate many of the household textiles they expected to take to their marriage, especially if their own family was already well equipped. A bed alone needed a mattress ticking made of coarse linen or cotton to contain the stuffing, two sheets of linen or cotton for summer and two of wool flannel for winter, cotton or linen pillow cases, several blankets (or in German areas, a linen-encased featherbed), and if possible, a coverlet. Describing the work of young back-woods girls in 1853, Samuel Strickland (brother to Catherine Parr Traill), drawing on his experiences of living in Canada for a quarter century, wrote:

When the wool is carded, it is ready for the wheel, and the ladies of the family, or if there are no ladies, their substitutes, hired spinning girls, convert it into yarn, which is then dyed, reeled, and hanked, which processes it undergoes at home. It is now ready for the weaver, who charges from fivepence to sixpence per yard for his work; many farmers, however, have hand looms in their own houses.

While some families made their own simple,

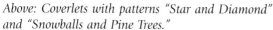

Above: Coverlets with patterns "Star and Diamond"
and "Snowballs and Pine Trees."
Right: A reversible coverlet
Below: Patchwork quilts were made from old clothing
and bedding.

household fabrics, it is possible to imagine a young girl daydreaming of the time when she would move with her new husband into her own home as she spun the yarn for a special coverlet to commemorate this transition. Overshot may have been the available choice of New Brunswick, Nova Scotia, and even some Quebec women, as early as the 1830s, but in Ontario the preference would have been for a jacquard coverlet. Not only did these represent the latest style, they served as important documents of special occasions, for the names or initials of the betrothed couple, for instance,

These simple, everyday homespun garments could have been made by any semi-skilled weaver. The plain checked design would not have required the expertise of a professional.

and perhaps a romantic symbol like a pair of hearts, could be woven into the corners. Though it cost extra to add the names — John Campbell charged $2.50 for weaving a coverlet, 50¢ for an extra colour, and 50¢ to weave a name, for which he had to cut special pattern cards — it was well worth it, since this one-time purchase would hold a place of honour in the new home. Part of what made commercial and fancy weaving like this viable was growing technology and new inventions.

Carding mills had dramatically expanded the raw material available for hand spinning early in the century, but the output of spinners was limited by the capacity of the wheels they used. Recognizing the importance of homespun to backwoods families, in 1865 the *Canadian Farmer* printed several proposals to increase production. For example, a letter in the March edition suggested the introduction of a two-handed wheel, used with some success in Europe, to double productivity. In June, Robert Brown wrote a letter to the editor hailing the advantages of a new invention, the "Improved Spinning Wheel" that adapted the walking wheel so a person could sit while

Above: Women could trade their homespun yarn or handwoven cloth at the local store for bolts of good-quality imported cloth or carpeting.
Left: A cradle, feather bed and trundle bed displaying various patterns used for bedding.

working. Such a device could expand the available work force by making it easier "for a person with one leg or one whom age has made comparatively infirm" to spin or to do so without tiring. George Potts, a commercial weaver, was so convinced that this contrivance was worthwhile that he paid $150 for the patent, and in 1866 added a workshop for its manufacture to his weaving business near Simcoe, Ontario. But the real change that would alter the home production of textiles was the new mills that expanded the earlier carding and fulling operations to include spinning and weaving.

4 From Farm to Factory

There was no specific moment in nineteenth-century Canada when factories took over the work of hand weavers. The process was gradual, beginning about 1810 with carding and fulling mills dotting the countryside, operated by a handful of workers. Small, integrated woollen mills proliferated mid-century and grew in number and size until the end of the 1800s, when urban cotton factories, employing hundreds of people, began to

Water for power came from a millpond that was channelled into a canal called a millrace that drove the turbines housed inside the mill building.

By the middle of the nineteenth century, small, integrated woollen mills, like the Asseltine Mill at Upper Canada Village, spun yarn and wove cloth from start to finish.

surpass wool production in both the quantity of cloth produced and the size of the labour force.

Instead of making homespun obsolete, the carding and fulling mills of the early nineteenth century actually expanded its manufacture. If we consider that the ratio of carders to spinners was six to one, it was still more cost-effective to pay for carding at the mill — it freed up children and women for other farm chores and made it possible to spin more. Once the needs of their families were met, girls might make money spinning for others, or by selling or trading extra yarn through the local store or to a market weaver. An 1817 report stated that in

Once the wool was washed, it was ready to be sent to the picking machine.

Ontario, for example, "A woman has from six to eight dollars per month for home work (house-work) and for spinning nearly as much." Many early farmers lived close to the line, and the family's well-being depended on products that women could trade or sell, such as butter, eggs, garden vegetables, maple sugar, yarn, and cloth. When times were tough, the ability to spin more yarn or weave more fabric could mean the difference between the hardship of poverty or comfortable survival.

Large families with lots of females had the best hedge against bad times. William Hutton of Ontario described how his wife and five daughters helped out the family economy early in the nineteenth century:

Anne is spinning hard for flannel frocks for herself and the girls … we have this day got home [from the weaver] 33 1/4 yards of grey cloth 3/4 wide [27 inches], worth here 6s.6d. per yard made from our own wool. Anne can

After preparation, the fleece was fed through a carding machine. This is a simple carding drum, similar to those found in the first carding mills.

spin 2 yards both warp and filling everyday and is become so good a spinner that we shall spin all our wool at home next summer and make 100 yards of cloth and 30 of flannel.

Although the women's work did not generate cash, nevertheless, says Hutton, "it is remunerating work in this way at least, that the cloth pays Tradesmen and Laborers and will possibly furnish us a carpet as the merchants will exchange excellent carpeting for homemade cloth yard for yard." The presence of a carding mill in almost every Canadian township made such production easier.

Many of the early carding mills combined fulling operations, the two processes most easily

As water-powered carding became more efficient, rather than the early single carding drum, a series of rollers guaranteed that the wool was evenly brushed so that it could be made into roving for the spinning mules.

adapted to water power. As mentioned earlier, prior to the establishment of fulling mills early in the nineteenth century, the French Canadians and Acadians accomplished the work by beating the cloth with mallets in a wooden trough while wet. The Scots also had a practice of manual fulling, though they did it differently. In Cape Breton Island, where the tradition survived, people sat around a long table that sometimes had grooves for draining the excess water. Everyone grabbed a section of wet cloth, pounded it back and forth, and then handed it on to the person to their left (the cloth was always passed in a clockwise direction, or "with the sun," as it was considered bad luck to go counterclockwise, or "against the sun"). To help lighten the hard, physical work, people made the occasion into a social get-together called a "waulking" or a "milling frolic," in which they sang special songs that told stories of life, love, and death. The music not only helped to pass the time entertainingly while maintaining a rhythm, the duration of each song also signalled the amount of shrink-

Above: The carded wool was wound onto large spools and fed through another machine to produce thin roving for the spinning mule.
Right: Roving ready to be spun

age. The blanketing or flannel that was finished in this manner was heavier and denser, hence warmer and more durable, than if it had simply been washed. But this was backbreaking work, and by the beginning of the nineteenth century in Canada, the same process could be accomplished using water-powered hammers to pound the cloth to shrink and clean it.

Carding and fulling machines were often housed in the same structure, situated on a river or stream, frequently owned by one man (though not all carding mills had fulling equip-

Water powered carding machines produced quantities of carded wool with evenly aligned fibres, needed for the automatic spinning machines. It was also used as stuffing for quilts.

ment). Some cloth, blanketing or coating for instance, had to undergo further processing, like the napping that brushed the wool to make it fluffier and warmer, important in a climate averaging only five degrees centigrade for six months of the year. Fuller's teasels (the heads of burr-like plants) were the natural brushes of choice for napping, even when the process was mechanized. Whereas carding and fulling mills represented the earliest use of water power to ease the housewife's lot, soon spinning and then weaving were brought under the same roof.

While there may have been one or two integrated mills that made cloth from start to finish by 1830, it is likely that, even if housed in a mill, the spinning and weaving were still done by hand. One of the earliest woollen factories in Lower Canada had only one small water-powered carding engine; a slubbing billy that turned carded wool into roving for spinning; a

The spinning mule required the operator to push the machine back and forth to twist the roving into yarn and wind it on the bobbins. This invention meant that hundreds of threads could be spun at once, instead of just one.

hand-operated spinning jenny with seventy-five spindles; and two hand looms. Requiring the labour of only two or three people, this was little more than an expanded professional weaver's workshop. By the 1850s, however, woollen mills with power spinning equipment and looms were well established, though still not large or numerous.

In 1851, Upper Canada had 74 woollen factories and 147 separate carding and fulling mills; in Lower Canada there were 18 factories and 193 carding and fulling establishments.

Most of the 81 mills in New Brunswick and 52 in Nova Scotia would have been for carding and spinning only. Though the interiors of the small factories could be dark, dusty, and noisy, most were run by one, and occasionally two, full-time workers, who might also have been the owners — a far cry from the "dark satanic mills" of England at the time. Though slightly larger, integrated woollen factories usually employed only 3 to 12 people, though one very large venture had 170. During this early phase of Canada's industrialization, unlike

Water-powered fulling stocks and fulled cloth. Fulling was one of the first textile processes to use water power to drive hammers that pounded the cloth to shrink it — much easier than doing it by hand.

England and the United States, the workforce was mostly male. Few Canadian women could be spared from the demanding routine of the farm in a country that was still predominantly rural. Interestingly, Ontario led the way in the industrializing process, a pattern for the future. Over the next two decades the number of Ontario woollen factories expanded, but carding and fulling establishments declined. This signified the replacement of handspun with mill-woven fabric, though in Quebec and the Maritimes many women continued to weave at home.

Early power looms could weave only basic

The spun and dyed yarn was wound in sections onto a warp beam that would be taken to the loom for weaving.

cloth, much like the homespun textiles of hand weavers. Coverlets or textiles with complex designs were beyond the capacity of these looms. For instance, in 1846, when James Rosamond of Lanark County, near Ottawa, made the full-scale conversion of his carding mill to a woollen factory, he bought wool from the local farmers and wove plain cloth that was the usual grey, though he could dye it as well. He also made cassimere (a textile woven in twill weave that did not need to be fulled), satinett (an inexpensive fabric made of cotton and wool), flannel, and blankets. As his prices dropped, it became cheaper for nearby farmers to buy his goods than to make their own. From this modest beginning, the business grew in 1866 to occupy a six-storey stone building containing a dye works, a boiler

Wool blanketing being woven on the power loom.

house, offices, and a warehouse, plus all the cloth-making equipment. By the end of the nineteenth century, the Rosamond Company had become Canada's largest woollen mill, with an international market.

Though early industrialization sustained home production of cloth for a while, over the course of the nineteenth century it devastated it, though not completely. With the exception of Quebec, Canadian linen production had practically disappeared by the 1830s as imported cotton warp yarn replaced the tedious work involved in flax processing. With the arrival of power spinning and weaving, Ontario farm families not only had a market for their wool, they could also purchase the blanketing and clothing material they once made at home. Some of the professional male weavers became the proprietors and operators of these mills. Recognizing the inevitable demise of their craft, others passed on their equipment to their less-skilled daughters, rather than train their sons. As the need for homespun declined and imported mass-

As hand weaving died out toward the end of the nineteenth century, some women continued to use hand looms to weave colourful rag rugs out of worn-out fabric.

produced cotton counterpanes made the wool and cotton coverlets seem old-fashioned, the demand for hand weaving narrowed to the production of things like rag rugs.

Catherine Parr Traill describes how thrifty households could recycle old cloth by cutting it into strips that were wound onto balls "as big as a baby's head" and used as the weft for carpeting. Once a housewife had accumulated enough rags,

she could commission one of the few remaining weavers to work them up. According to Traill,

a pound and a half of rags will make one yard of carpet.... Bits of bright red flannel, of blue, green, or pink mousselin-de-laine, or stuffs of any bright colour: old shawls and handkerchiefs, and green baize will give you a good, long-enduring fabric, that will last

This device maintained the tension and sequence of each thread as it is wound on the warp beam.

for eight or ten years. Narrow widths could be used for stair carpeting; sewn together in strips, they could carpet a room wall to wall.

Rural Quebec farm families wove rag rugs too, but they also continued to grow linen and wool and weave simple fabrics into the twentieth century, especially those who lived some distance from Quebec City or Montreal. While Ontario carding and fulling mills declined between 1851 and 1891, they actually increased in Quebec — an indication that domestic spinners and weavers still relied on their services. The distance from cities, combined with a lack

of cash, made cotton warp and factory-made cloth hard to come by. Large families, however, provided enough labour for them to continue making their own cloth — mostly wool and flannel, and in some areas a surprising amount of linen, given its demise elsewhere.

New Brunswick farm women continued to make homespun too, but for profit. The lumber industry provided a market for their cloth because it was warmer and harder-wearing than factory goods. As late as 1880, merchants advertised for "home spun cloth of all kinds in large quantities, also 2000 pairs of socks and mitts, 1-2 ton woolen yarn, over socks, home

The lightly twisted roving was wound onto huge spools and stored on racks, ready to be spun.

knit drawers, shirts, pants etc." In exchange, the women could choose from a large array of dry goods, including the latest dress fabrics. In other parts of the Maritimes, Scots and Acadians continued to hand weave well into the twentieth century, for the same reasons as in Quebec — isolation and necessity. Mrs. John P. Monroe, the last professional overshot coverlet weaver on Cape Breton Island, for example, made over one hundred coverlets in her lifetime and continued to weave until her death in 1971. For much of the nineteenth century, hand spinning and weaving co-existed with industrial production. Eventually, however, the big mills took over most of the work, and in the twentieth century, hand cloth production became a rarity.

Epilogue

The transformation from farm to factory affected professional and domestic weavers unevenly and regionally, though it never entirely eliminated hand production. Canada was urbanizing throughout the period, but it was still a large, relatively empty country. Even as the twentieth century dawned, some rural areas in eastern Canada continued to produce homespun and the newly opened western settlements attracted immigrants such as the Ukrainians, Icelanders, and Doukhobors, who re-established some textile traditions in their new homes.

Spinning and weaving in Canada didn't completely disappear in the early twentieth century, despite a growing textile industry. Not only did women in isolated regions of the Maritimes and Quebec continue to use their wheels and looms, but as the railway opened up the West for settlement, immigrants like

A colourful Ukrainian bench cover, woven with wool and dyed with twentieth-century commercial dyes.

The custom of spinning frolics, where groups of women gathered with their wheels to spin the yarn needed for a project, continued on Cape Breton Island into the early twentieth century.

the Ukrainians, Icelanders, and Doukhobors found that like the nineteenth-century Eastern pioneers, they needed to use their cloth-making skills for survival and to get ahead.

The Acadians in Cape Breton Island, for example, continued to weave simple, functional textiles. To make their rag rugs, they purchased colourful yellow, red, and green carpet warp instead of the more readily available blue and white warp yarn. Though they still used some vegetable dyes, store-bought synthetic dyes made it easier to create the women's pink homespun petticoats and the men's red long underwear that brightened the hard lives of Acadians dwelling on the rocky shores of the Atlantic. Living close to the line, some people reportedly eked out scarce wool by carding it with winter-white rabbit fur (four parts wool to one of fur) and by combining rags and yarn to form attractively banded patterns for simple blankets and bed covers. The Scots of Cape Breton continued their spinning frolics, where women carried their wheels across the fields to their host's house in the

The design of this French Canadian coverlet is created by pulling an extra weft thread into loops.

morning, spun all day, and were joined by the men for dinner and an evening of music, singing, and dancing. In addition to basic homespun, female (and some male) weavers made shepherd's plaid (usually a brown-and-white check), tartan, and overshot coverlets with pattern names like "The Forsaken Lover," "Carts and Wheels," and "Bachelor Among the Girls" well into the twentieth century.

In parts of rural Quebec, in addition to the indispensable utilitarian blankets, towels, and grain sacks, some weavers made distinctive bed coverings that were simply constructed but imaginatively designed. Although originally inspired by nineteenth-century, fashionable all-white, tufted cotton counterpanes imported from Britain, the French Canadian imitations evolved into exuberant and distinctive multi-coloured textiles. Using a mixture of cotton, linen, wool, and sometimes even rags, the weaver created a pattern using an extra weft of colourful wool yarn, which she drew into a

The Ukrainians who settled the Prairies brought their textiles traditions with them, including the long bench covers and woven tapestry wall hangings.

loop, a technique known in Quebec as *bouton-né*. Over time, the formal structure of the British examples, usually with an eight-pointed star in the middle, bordered by pine trees and flowers, gave way to a free-form use of these motifs that was unique to Quebec. These patterns provided a creative outlet for the nineteenth-century weavers, but by the 1920s, the technique and designs had become more standardized in attempts to promote Quebec handicrafts to a wider market.

As the spinning and weaving activities that had been such an important part of the nineteenth-century household economy in eastern Canada yielded largely to department stores and mail-order catalogues in the twentieth, Western Canada was being newly developed. Although the English settlers had long since abandoned their wheels and looms, Icelandic, Ukrainian, and Russian settlers had not. The spinning wheel or spindle was an especially useful tool for some of these new immigrants as they began life in a new country, and in some cases weaving too played a part.

The Icelandic settlers, who arrived in Manitoba in 1870, raised sheep and spun yarn on space-saving upright wheels. This yarn was used for knitting underwear, socks, mitts, and sweaters for their families and for sale. By 1891, some communities were doing remarkably well, according to a report of the North-West Mounted Police, which commented that "they are very industrious, and I noticed all the women knitting, even while driving their cattle." As in the eastern Canadian frontier settlements, the work of the women was critical to a successful family economy. Along with the Icelandic settlers, two other groups that carried on their distinctive textile traditions in Canada were the Ukrainians and the Doukhobors. With their origins in Eastern Europe and the Caucasus, they brought techniques, colours, and designs for their fabrics that were as strikingly unlike those of eastern Canadians as were their furnishings and furnished textiles.

Ukrainians arrived in Canada in numbers during the 1890s in response to a campaign by the government to attract agricultural workers from Eastern Europe to settle the West. Their early buildings resembled those they left behind, and the great room of the house was the showcase for some of the spectacular textiles that created the feeling of home. Here was placed the long dining table, draped with a crisp white linen tablecloth, and a seating bench with a cover woven of hemp, wool, and, later, rags banded in vibrant yellows, greens, blues, and reds. Along the top of the wall, there might be a long woollen, tapestry woven in traditional geometric shapes. One woman recalled that her Ukrainian immigrant mother "worked extremely hard in the field and the home, with no modern conveniences but always found time for weaving, knitting, crocheting, and embroidery ... her flowers bloomed among the vegetables in the summer

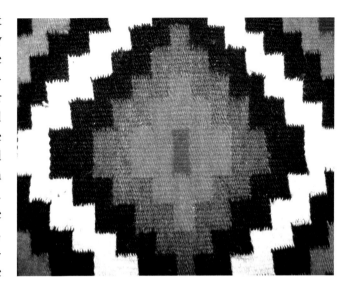

Top: The details of a wall hanging show clearly the tapestry weaving technique.
Above: The vibrant purples, oranges, pinks, and greens may have been dyed using commercial Diamond dyes from Montreal.

and on her floor and furniture in the winter."

Another group that arrived at about the

When not in use, clothing and household textiles were folded and stored in a wardrobe.

same time was the Doukhobors. A pacifist religious sect from the Caucasus, they settled first in Saskatchewan, but some moved on to British Columbia, where they lived communally in buildings that housed thirty to fifty people. According to textile scholar Dorothy Burnham, most women had their own spinning wheels, perhaps received as a special gift on their thirteenth birthday, on which they spun linen and wool for colourful clothing, household textiles, and carpets — but they shared the single loom in the house. Although

they were largely self-sufficient, the Doukhobors used packaged dyes from Montreal to obtain the vivid pinks, greens, purples, yellows, and blacks so common in their textiles. Beds did double duty. They served as sleeping shelves at night and as seating during the day, when the bedding, folded at one end, was concealed under a bedding cover woven of wool cloth decorated with elaborately knitted multicoloured borders. The women's skirts and aprons had similar decorative edges. For their floors, the Doukhobors made tapestry and knotted-pile carpets on vertical looms, using patterns typical of the rug-weaving region from which they came. The textile work of the Doukhobor women, like that of all European settlers since the Vikings, was an essential part of survival in a newly developing country.

In our throw-away world, it is difficult to imagine the value of cloth to nineteenth- and early twentieth-century Canadians. Time-consuming to make and expensive to buy, textiles were used and reused until there was nothing left. Worn-out adult clothing was cut down for children, and when that was too threadbare to be useful, it might be stuffed in the chinks of a log cabin to keep the wind out. Old bedding and textile scraps were saved to make rag rugs. Quilts, with their designs pieced together from small remnants of fabric, are often the only clues we have left of the everyday cloth produced out of necessity and pride by wave upon wave of immigrants; cloth that tells the story of necessity and comfort, of economy, and of getting ahead in a frontier society.

Glossary

baize: a coarse woollen fabric with a long nap, used for such things as curtains, petticoats, shirtings, and linings.

bird's-eye twill: a fabric woven in twill weave to form a small lozenge-shaped pattern with a dot in the centre so that it resembles a bird's eye.

bobbin: the spool on a spinning wheel which held the spun yarn. Also a cylinder, usually of wood, onto which the weft yarn was wound (also called a "quill"). When filled, it was inserted into the shuttle so that the weft unwound as it crossed the warp.

bobbin winder or quill winder: a tool used to wind yarn onto bobbins.

boutonné: a weave structure in which an extra weft is pulled up to make a loop in a sequence that forms a pattern. Complex designs can be created with this technique, even using very simple looms.

calico: a cloth made from cotton that first came from India and was later made in the west. It came in many qualities and varieties and was used for both men's and women's clothing

cambric: originally a fine white linen cloth made in Cambray in Flanders; by the nineteenth century it was also made from cotton and was used for things like shirt ruffles or trimming.

capot: a coat that resembles a frock coat with a hood. In French Canada this was usually made of grey homespun and was worn with a colourful woven sash.

cards/carding: a pair of brush-like devices that usually consisted of bent wires mounted on a small slab of wood with handles; used to brush raw wool to remove the tangles and align the fibres so they could be spun into a consistent yarn without tangles or knots. The action involved in using cards is called "carding".

cassimere: a woven twill weave using a fine yarn. Unlike broadcloth, it did not need to be fulled (though some was).

click-reel/clock reel: a tool used to wind spun yarn off the bobbin of the spinning wheel and measure it (critical information if the yarn was to be used for weaving). A clock-like dial measured the revolutions the reel had made, and a loud click alerted the operator when a designated yardage had been reached.

cloth beam: a beam at the front of the loom onto which the cloth was wound once woven.

country cloth: a term often used interchangeably with "homespun" to designate a simple, coarse, hard-wearing cloth. In French the term is *étoffe du pays*. See also "homespun."

couvertures de mariage: a term used by the Acadians to designate an all-white bridal coverlet, often made of recycled cloth, that was part of a bride's dowry

distaff: a holder used to keep long flax fibres from tangling during spinning. It could be part of a spinning wheel or be carried under the arm to free both hands for spinning.

étoffe du pays: see "country cloth" and "homespun"

felting: a process (usually washing and pounding) that allows wool fibres to be matted together to form a warm, dense fabric (see also "fulling")

flax break: a wooden implement on a stand with a top and a bottom that are hinged together. The stalks of flax are laid over the bottom piece and the top blade is forcefully brought down onto the flax repeatedly so as to break and remove the outer bark, making the inner fibre accessible.

fly shuttle: a spring-loaded device that mechanically shot the shuttle back and forth across the warp so the weaver did not have to throw it and catch it.

flyer and bobbin: the mechanism of a spinning wheel that simultaneously spins the yarn and winds it (the flyer) onto a bobbin. It allows a spinner to use both hands to spin a fine, consistent yarn.

fulling: a process whereby woven woollen cloth is washed and shrunk to make it thicker and warmer. It could be done manually by pounding it with mallets or by hand, or in a water-powered mill using hammers.

great wheels: spinning wheels that had a large wheel (approximately thirty-six inches in diameter) turned by hand, with a drive belt attached to a spindle turned on its side. The spinner walked back and forth to spin the yarn and wind it on the spindle. Also called a "walking wheel."

Guernsey frock: a thick, warm, knitted sweater or vest that fit snugly.

hackles/ hackling: tools with long metal prongs set in wood used to comb flax fibres.

homespun: a coarse, hard-wearing fabric, usually made of wool or a combination of wool and linen or cotton. It is simple to make and does not require the skills of a professional weaver.

jacquard loom: a loom that uses a series of punched cards to manipulate individual warp ends and make a realistic pattern.

linsey-woolsey: a fabric woven in plain weave with a linen (later cotton) warp and wool weft. It was woven domestically and imported from Europe to be used in clothing like aprons and petticoats, skirts, and men's work clothes. It could also be used for bedding.

maidens: the two upright posts that held the flyer and bobbin of a spinning wheel.

mantelet: a short, loose cape that covers the shoulders.

market weavers: weavers who sell some or all of the cloth they make.

marriage coverlets: see *couvertures de mariages*

mordant: a substance that combines with a dye to allow the colour to become fixed.

mother-of-all: the name for the part of the spinning wheel that contains the maidens, flyer, and bobbin.

mousselin-de-laine: a fine dress material printed with varied patterns, originally made entirely of wool, but later chiefly of wool and cotton.

muslin: a lightweight cotton fabric

nap/napping: the fuzzy surface of a woollen cloth, made by brushing it (napping)

niddy-noddy: a simple hand-held reel used to remove the spun yarn from the spinning wheel and make a skein.

overshot: a weave structure in which an extra weft yarn "shoots over" a group of warp ends to form a pattern. It was most often used in bed coverlets, but sometimes in table linens.

petite étoffe: a French term for flannel or homespun.

plain weave: see "tabby"

quill wheel: see "bobbin winder"

rippling: the process of removing the seeds from a flax plant.

roving: a long cylinder of unspun carded wool produced with water-powered carding machines.

satinett: a fabric woven of cotton and wool with a smooth satin-like finish.

shepherd's plaid: an all-wool fabric in a black or dark blue and white check — probably in houndstooth pattern.

singles: when yarn is spun, it forms a single strand ("singles"); two or three strands can be plyed together (two-ply or three-ply yarn), often for use in knitting.

skutching knife: a wooden device with a sharp side, used after breaking the flax to remove the broken bark from the fibre (see flax brake).

slubbing billy: an early mechanized device that transformed machine-carded wool into a lightly spun yarn that was strong enough to withstand the mechanical spinning process.

spindle and whorl: a simple device used to spin yarn, made with a stick or rod weighted at one end with a whorl. Twisted with the thumb and forefinger it can be supported in some way or dropped to spin freely. It can be turned on its side and driven by a wheel to speed up the process.

stuff: a generic term used to describe a woven fabric, though usually it refers to a fabric woven without a nap or pile.

swift: a device that holds a skein of yarn as it is being wound into a ball for knitting or onto spools for making a warp or bobbins for weaving.

tabby: also called "plain weave," it is the most basic weave structure, in which the weft yarn passes over and under alternate warp ends.

twill: a weave structure in which the weft passes over one and under two or more warp ends (instead of over and under in regular succession, as in plain weaving), creating slightly raised parallel lines.

twill diaper: a complex twill weave construction that produces a large overall pattern that is heightened by the use of two colours.

walking wheel: see "great wheel"

warp: the lengthwise threads of a woven cloth; each strand of warp is called an "end" (see also "weft").

warping board/warping mill: devices on which warp threads can be measured and kept parallel to put on the loom. It can be a simple frame with pegs which is mounted on or leaned against a wall, or a revolving mill.

waulking: a term used in Scotland (and Cape Breton Island) for fulling the woven woollen cloth (see "fulling"). Also called "milling."

weft: the crosswise threads of a woven cloth. A single weft is called a "pick." Each weft passes the opening caused by the sequential raising of some warp ends and the lowering of others that allows a shuttle to carry the weft thread through.

Sites of Interest

NEWFOUNDLAND

L'Anse aux Meadows National Historic Site
Early 1100s

P.O. Box 70
St-Lunaire-Griquet, NL, A0K 2X0
Phone: 709-623-2608
viking.lam@pc.gc.ca
www.pc.gc.ca/lhn-nhs/nl/meadows/index_e.asp

An archaeological site of the earliest known European settlement in the New World. Exhibits, reconstructions and re-enactments highlight the Viking lifestyle of the early 1100s, including domestic technologies such as cooking and textile making.

NEW BRUNSWICK

Kings Landing Historical Settlement
1820 to 1909

20 Kings Landing Road,
Kings Landing, NB, E6K 3W3
Phone: 506- 363-4999
www.kingslanding.nb.ca/

An outdoor museum that recreates rural, nineteenth-century, Loyalist New Brunswick. Textile production is interpreted and special Heritage Workshops provide a hands-on experience of spinning and weaving.

Village Historique Acadien
1770 to 1939

C.P. 5626
Caraquet, NB, E1W 1B7
Phone: 1-877-721-2200 or 506-726-2600
vha@gnb.ca
www.villagehistoriqueacadien.com/main.htm

An award-winning historic site with over forty complexes. Both flax and wool production is interpreted through all stages of preparation.

NOVA SCOTIA

Barrington Woollen Mill
Barrington Woolen Mill Museum
2368 Highway 3
Barrington, NS, B0W 1E0
Phone: 902-637-2185
http://museum.gov.ns.ca/bwm/index.htm

This site represents an integrated turbine-driven woollen mill, established in 1882 as a community enterprise. Its carding machines, spinning mule, loom twister, and skeiner remain in their original places, as does the water turbine that powered the machinery. There are also exhibits on sheep raising and wool processing in the area, and demonstrations of hand spinning.

Historic Sherbrooke Village
1860 to pre-World War I

P.O. Box 295
Sherbrooke, NS, B0J 3C0
Phone: 1-888-743-7845 or 902-522-2400
svillage@gov.ns.ca
http://museum.gov.ns.ca/sv/index.php

An historic village recreated to interpret many activities from the second half of the nineteenth century. McMillan House, originally built prior to 1870 by Dan McMillan, is now recreated as a weaver's residence. Textile demonstrations are done in the kitchen, and looms are set up in the back room. There is also a tailor's shop restored to 1900.

Nova Scotia Highland Museum
4119 Highway 223
Iona, NS, B2C 1A3
Phone: 1-866-4GAELIC (1-866-442-3542) or 902-725-2272
highlandvillage@gov.ns.ca
http://museum.gov.ns.ca/hv/index.html

A living history museum, on a forty-three-acre hillside overlooking the world-renowned Bras d'Or Lake. It interprets Nova Scotia's rich Scottish Gaelic culture with costumed staff, farm animals, and period buildings and artifacts. There is a small wool-carding mill, and spinning, weaving, and cloth milling (waulking or fulling) are interpreted.

Ross Farm Museum
Mid 1800s

R.R. # 2, 4568 Highway #12
New Ross, NS, B0J 2M0
Phone: 1-877-689-2210
http://museum.gov.ns.ca/rfm/

Once the home of five generations of the Ross family, the farm was first cleared from the forest in 1816. The museum shows how the early settlers and their descendants lived and coped with the land around them. Wool spinning is interpreted, and there is a Heritage Animal Program that breeds South Down and Cotswold Sheep.

Wile Carding Mill Museum
P.O. Box 353, 242 Victoria Road
Bridgewater, NS, B4N 2W9

Mailing Address:
c/o deBrisay Museum,
60 Pleasant Stree
Bridgewater, NS, B4V 3X9
Phone: 902-543-8233
http://manl.nf.ca/atlantic/wilecarding.html or
http://museum.gov.ns.ca/wcm/index.htm

Costumed interpreters tell of working life in the noisy, damp, water-powered carding mill while demonstrating wool picking, hand-carding, tabletop machine carding, and drop-spindle spinning. Demonstrations of sheep shearing and wool processing are performed on-site for special events.

QUEBEC

The Ulverton Woolen Mills
Mid 1800s

210, chemin Porter
Ulverton, PQ, J0B 2B0
Phone: 819-826-3157
info@moulin.ca
www.moulin.ca/PagesAng/home.htm

The original integrated textile mill situated on the Ulverton River was built in 1850. It now interprets the pre-industrial and industrial production of woollen fibres and threads. The historic machines, many more than a century old, are fully functional. There are four floors of thematic exhibitions related to wool and textile production.

Village Québécois d'Antan
1810 to 1910

1425, Montplaisir Street
Drummondville, PQ, J2B 7T5
Phone: 1-877-710-0267 or 819-478-1441
www.villagequebecois.com/pages/ang/accueil.htm

An historic site containing authentic and reproduction buildings, it has a carding mill (c. 1885); the Morel House (c. 1890), where a woman weaves typical Québécois rugs; the peasant-style Houle House (c. 1827), where a woman spins wool; and the Blanchette-Vallières House (c. 1870), where linen is woven. There is also an Abenaki settlement with basket-making.

ONTARIO

Black Creek Pioneer Village
1860s

1000 Murray Ross Parkway
Toronto, ON, M3J 2P3
Phone: 416-736-1733
bcpvinfo@trca.on.ca
www.blackcreek.ca/

The site has over thirty-five authentic buildings to recreate a village of the 1860s. Charles Irving's professional weave shop has a fly-shuttle hand loom, a rag loom, a warping mill, and various tools used by a professional weaver, including a spool rack and spool winder and a quill winder to fill the bobbins for the shuttles. Spinning and dyeing are interpreted at the Strong farmstead (dating to 1815) and sheep graze in a pasture confined by a split-rail fence.

Joseph Schneider Haus Museum
1816

466 Queen Street South
Kitchener, ON, N2G 1W7
Phone: 519-742-7752
bamarie@region.waterloo.on.ca
www.region.waterloo.on.ca (follow the links through "visiting" to "museums.")

A small complex, consisting of a small exhibit space and a fine Georgian frame farmhouse built by one of the area's first pioneers, Joseph Schneider, a Pennsylvania–German Mennonite. The house is Kitchener's oldest dwelling (c. 1816) and is restored and furnished to period. The beds are made up with spectacular handwoven linens and beautiful coverlets. The collections contain an impressive array of handmade textiles and tools that are periodically the subject of special exhibits.

Lang Pioneer Village

1820 to 1899

104 Lang Road
Keene, ON
Mailing Address:
County of Peterborough
Attn: Lang Pioneer Village Museum
470 Water Street
Peterborough, ON, K9H 3M3
Phone: 1-866-289-5264 or
705-295-6694
http://langpioneervillage.ca/

The site consists of twenty restored and furnished buildings constructed between 1820 and 1899. Spinning and vegetable dyeing are interpreted. Of particular interest is the jacquard loom that belonged to commercial weaver Samuel Lowry, who wove in the Peterborough area in the late nineteenth century. The loom is in the process of being restored, and the site is currently in the planning stages of a new "Weaver's Shop," which will be built on-site to house the loom and is scheduled to open in the near future. Check with the site for details.

Mississippi Valley Textile Museum

1867 to present

3 Rosamund Street East, Box 784
Almonte, ON, K0A 1A0
Phone: 613-256-3754
www.textilemuseum.mississippimills.com

Located in the annex of the former Rosamond Woolen Company, built in 1867 and now featuring a blend of the old and new history of the Mississippi Valley and the textile industry. Exhibits range from early mill history and period mill equipment to cottage-industry and eclectic modern fibre-art exhibitions. This community museum combines traditional static and working displays of textile equipment and processes, with activities and events focusing on the region's heritage and culture and the role of the textile industry in the development of Canada.

Ontario Science Centre

770 Don Mills Road
Toronto, ON, M3C 1T3
Phone: 416-696-1000
webmaster@osc.on.ca
www.ontariosciencecentre.ca

The Science Centre has an operating jacquard loom that belonged to Scottish-born weaver John Campbell, who worked near London, Ontario, in the mid-nineteenth century. Located in the Communication Hall on Level D, the loom is set up to demonstrate what a 150-year-old device that uses punch cards to produce a pattern has in common with modern computers. The loom is operated once a week, usually on Wednesdays from 11:00 a.m. to 3:00 p.m.

Peterborough Centennial Museum

Museum Drive at Hunter Street East (Armour Hill), P.O. Box 143
Peterborough, ON, K9J 6Y5
Phone: 705-743-5180
www.pcma.ca/index.htm

It has an excellent collection of nineteenth-century Ontario homespun and a good quilt collection. These are not on permanent display; some of the material is available as a virtual exhibit. Check website for details.

Textile Museum of Canada

55 Centre Avenue
Toronto, ON, M5G 2H5
Phone: 416-599-5321
info@textilemuseum.ca
www.textilemuseum.ca

The permanent collection contains more than twelve thousand textiles and spans almost two thousand years and two hundred world regions. The website has a searchable database where it is possible to find Canadian hand-woven material. There is a varied and frequently changing exhibition schedule.

Upper Canada Village

1860s

13740 County Road 2
Morrisburg, ON, K0C 1X0
Phone: 1-800-437-2233 or 613-543-4328
www.uppercanadavillage.com

A dynamic historic site that interprets daily life in a small 1860s village. The McDiarmid House interprets wool and flax spinning, vegetable dyeing, and hand weaving. The Asseltine Woollen Mill (built 1840–41) is one of the few North American nineteenth-century water-powered textile mills that is fully operational, including mechanical carding, spinning, and cloth finishing. Blankets woven on the power loom with wool from the site's sheep are for sale in the gift shop. There is also a dressmaker's shop and a good general store showing the kinds of imported fabrics used at the time.

Westfield Heritage Village
1800s

1049 Kirkwall Road
Rockton, ON, L0R 1X0
Phone: 1-800-883-0104 or 519-621-8851
westfield@speedway.ca
www.westfieldheritage.ca

A living-history site of over thirty buildings. The Marr Shop, a small (c.1800s) timber-frame building, features demonstrations of spinning and weaving using a variety of spinning wheels and looms. Of particular interest is the unique spinning-wheel shop, with much of its original equipment. It was owned by George Potts, a farmer and weaver who built it on his farmstead near Simcoe in 1866 after buying the patent rights to make a new and improved walking wheel.

MANITOBA

Mennonite Heritage Village
early 1900s

231 PTH 12 N.
Steinbach, MB, R5G 1T8
Phone: 1-866-280-8741 or 204-326-9661
info@mennoniteheritagevillage.com
www.mennoniteheritagevillage.com

Interprets the Mennonite way of life from the sixteenth century to the present day. The forty-acre (seventeen-hectare) site spreads out from a village street, in a pattern reminiscent of Mennonite villages found throughout southern Manitoba at the turn of the twentieth century. The north side of the street illustrates the early settlement buildings, while the south side shows the gradual shift to various business enterprises. Sheep shearing, wool spinning, and quilting are demonstrated.

ALBERTA

Ukrainian Cultural Heritage Village
Early 1900s

8820–112 Street
Edmonton, AB, T6G 2P8
Phone: 780-662-3640
http://tapor.ualberta.ca/heritagevillage/

An open-air museum, built to resemble pioneer settlements in east-central Alberta. Buildings from the surrounding communities have been moved to the village and restored to various years within the first part of the twentieth century. Colourful household textiles are used in a variety of buildings.

BRITISH COLUMBIA

Doukhobor Discovery Centre (formerly Doukhobor Village Museum)
1908 to 1938

112 Heritage Way
Castlegar, BC, V1N 4M5
Phone: 250-365-5327
larryewashen@telus.net
http://doukhobor-museum.org/

A reconstructed communal village that introduces Doukhobor culture and its unique lifestyle as it evolved in the Kootenay region of British Columbia from 1908 to 1938. On display are naturally dyed fabrics and clothing made from home-manufactured linen, hemp, and wool, as well as tools and implements.

Notes

Below are the sources of quotations reprinted in the text and information gleaned from the works of other scholars. They are listed in the order in which they appear. For complete information on the sources, see the Selected Bibliography.

CHAPTER 1

"These French Acadians are hard-working" — John Mack Faragher, *A Great and Noble Scheme: The Tragic Story of the Expulsion of the French Acadians from Their American Homeland*. New York: W.W. Norton & Co., 2005, p. 183; For example, in 1791, Joseph Blanchard and his wife — Allan Greer, *Peasant, Lord, and Merchant: Rural Society in Three Quebec Parishes, 1740-1840*. Toronto: University of Toronto Press, 1985, p. 34; "the grey capote of the habitant" — Joseph Bouchette, *The British Dominions in North America, or, A Topographical and Statistical Description of the Provinces of Lower and Upper Canada, New Brunswick, Nova Scotia, the Islands of Newfoundland, Prince Edward, and Cape Breton Including Considerations on Land-Granting and Emigration: to Which Are Annexed, Statistical Tables and Tables of Distances, &c.* London: H. Colburn and R. Bentley, 1831, pp 406–07; All the fabrics discussed above — Dorothy K. Burnham, *Warp and Weft: A Textile Terminology*. Toronto, Ontario: Royal Ontario Museum, 1980.

CHAPTER 2

"it had been part of the marriage outfit" — Margaret M. Robertson, *Shenac's Work at Home: A Story of Canadian Life*. London: Religious Tract Society, 1868, pp. 50–51; "in June came sheep-washing" — Canniff Haight, *Country Life in Canada Fifty Years Ago: Personal Recollections and Reminiscences of a Sexagenarian*. Toronto: Hunter, Rose, 1885, p. 45; "all dirty wool is thrown aside" — Catherine Parr Traill, *The Canadian Emigrant Housekeeper's Guide*. Toronto: Lovell & Gibson, 1862, p. 109; "the hum of the spinning wheel" — Haight, *Country Life in Canada*, p. 46; "to enable him to extend his Manufactories" — Journal and proceedings of the House of Assembly, 1827. Halifax, N.S. s.n., 1827, p. 167; "men's socks sell at one shilling and six pence" — Traill, *The Canadian Emigrant Housekeeper's Guide*, p. 116; "her dress is home-spun" — Emily Elizabeth Beaven, *Sketches and Tales Illustrative of Life in the Backwoods of New Brunswick, North America: Gleaned from Actual Observation and Experience during a Residence of Seven Years in That Interesting Colony*. London: G. Routledge, 1845, p. 29.

CHAPTER 3

"it was full as cheap to go to the store" — Gourlay, *Statistical Account of Upper Canada*, p. 439; Luxury fabrics such as silks — Douglas McCalla, "Textile Purchases by Some Ordinary Upper Canadians, 1808-1861." *Material History Review* 53 (Spring-Summer, 2001), p. 53; "the state bedroom as well" — Beaven, *Sketches and Tales*, pp. 30–31; "respectfully informs the Inhabitants" — Harold B. Burnham, and Dorothy K. Burnham, *Keep Me Warm One Night: Early Handweaving in Eastern Canada*. Toronto: University of Toronto Press in co-operation with the Royal Ontario Museum, 1972, p. 12; "When the wool is carded" — Samuel Strickland and Agnes Strickland, *Twenty-Seven Years in Canada West, or, The Experience of an Early Settler*. London: R. Bentley, 1853, p. 293; Though it cost extra to add the names — Burnham and Burnham, *Keep Me Warm One Night*, p. 321.

CHAPTER 4

Small, integrated woollen mills proliferated — A.B. McCullough, *The Primary Textile Industry in Canada: History and Heritage.* Studies in Archaeology, Architecture and History. Ottawa: National Historic Sites, Parks Service, Environment Canada, 1992, p. 51; "A woman has from six to eight dollars" — Robert Gourlay, *Statistical Account of Upper Canada: Compiled with a View to a Grand System of Emigration.* London: Simpkin & Marshall , 1822, p. 439; When times were tough, the ability to spin — Janine Roelens and Kris Inwood, "'Labouring at the Loom': A Case Study of Rural Manufacturing in Leeds County, Ontario, 1870." *Canadian Papers in Rural History,* Volume VII, edited by Donald H. Akenson. Gananoque, Ontario: Langdale Press, 1989, p. 216; "Anne is spinning hard for flannel frocks" — Marjorie Griffin Cohen, *Women's Work, Markets, and Economic Development in Nineteenth-Century Ontario.* Toronto: University of Toronto Press, 1988, p. 79; "One of the earliest woollen factories" — McCullough, *The Primary Textile Industry,* p. 49; For instance, in 1846, when James Rosamond — Harold D. Kalman, *A Textile Museum for Almonte: Phase 1: The Feasibility of Establishing a Textile Museum.* Ottawa: National Capital Commission, 1983, pp. 10, 12. Printed courtesy of the Commission; "a pound and a half of rags" — Traill, *The Canadian Emigrant Housekeeper's Guide,* p. 113; Rural Quebec farm families wove rag rugs too — Sophie-Laurence Lamontagne and Fernand Harvey, *La Production Textile Domestique Au Québec, 1827–1941, Une Approche Quantitative et Régionale.* Transformation Series 7. Ottawa: Musée national des sciences et de la technologie, 1997, p. 15; "home spun cloth of all kinds" — Béatrice Craig, Judith Rygiel, and Elizabeth Turcotte, "The Homespun Paradox: Market-Oriented Production of Cloth in Eastern Canada in the Nineteenth Century." *Agricultural History* 76, no. 1 (2002), p. 49; Mrs. John P. Monroe, the last professional overshot coverlet weaver — Florence MacDonald Mackley, *Handweaving in Cape Breton.* Sydney, N.S.: privately printed, 1967, pp.28–31.

EPILOGUE

The Acadians in Cape Breton Island, for example — Mackley, *Handweaving in Cape Breton,* pp. 87, 20; In parts of rural Quebec, in addition to the indispensable utilitarian blankets — Adrienne D. Hood and David-Thiery Ruddel, "Artifacts and Documents in the History of Quebec Textiles," in *Living in a Material World: Canadian and American Approaches to Material Culture,* edited by Gerald L. Pocius. St. John's, Newfoundland: Institute of Social and Economic Research, 1991, pp. 76-88; The Icelandic settlers — Dorothy K. Burnham, *The Comfortable Arts: Traditional Spinning and Weaving in Canada.* Ottawa: National Gallery of Canada, National Museums of Canada, 1981, p. 205; "They are very industrious" — "Report of the Commissioner of the North-West Mounted Police, 1890," in Sessional papers of the Dominion of Canada: volume 15, first session of the seventh Parliament, session 1891, Ottawa: B. Chamberlin, 1891. p. 54; "worked extremely hard in the field and the home" — Burnham, *The Comfortable Arts,* p. 216; Most women had their own spinning wheels — Dorothy K. Burnham, *Unlike the Lilies: Doukhobor Textile Traditions in Canada.* Toronto: Royal Ontario Museum, 1986.

Selected Bibliography

Arsenault, Jeanne. "A la recherche du costume Acadian." *Material History Bulletin* 4 (1977): 46–56.

Beaven, Emily Elizabeth. *Sketches and Tales Illustrative of Life in the Backwoods of New Brunswick, North America: Gleaned from Actual Observation and Experience during a Residence of Seven Years in That Interesting Colony.* London: G. Routledge, 1845.

Bouchette, Joseph. *The British Dominions in North America, or, A Topographical and Statistical Description of the Provinces of Lower and Upper Canada, New Brunswick, Nova Scotia, the Islands of Newfoundland, Prince Edward, and Cape Breton Including Considerations on Land-Granting and Emigration: to Which Are Annexed, Statistical Tables and Tables of Distances, &c.* London: H. Colburn and R. Bentley, 1831.

Burke, Susan M. "Perpetuation and Adaptation: The Germanic Textiles of Waterloo County, 1800–1900," in *From Pennsylvania to Waterloo: Pennsylvania Folk Culture in Transition*, edited by Susan M. Burke and Mathew H. Hill. Kitchener, Ontario: Friends of the Joseph Schneider Haus, 1991.

Burnham, Dorothy K. *The Comfortable Arts: Traditional Spinning and Weaving in Canada.* Ottawa: National Gallery of Canada, National Museums of Canada, 1981.

_____. *Unlike the Lilies: Doukhobor Textile Traditions in Canada.* Toronto: Royal Ontario Museum, 1986.

_____. *Warp and Weft: A Textile Terminology.* Toronto: Royal Ontario Museum, 1980.

Burnham, Harold B., and Dorothy K. Burnham. *Keep Me Warm One Night: Early Handweaving in Eastern Canada.* Toronto: University of Toronto Press in co-operation with the Royal Ontario Museum, 1972.

Cohen, Marjorie Griffin. *Women's Work, Markets, and Economic Development in Nineteenth-Century Ontario.* Toronto: University of Toronto Press, 1988.

Craig, Béatrice, Judith Rygiel, and Elizabeth Turcotte. "The Homespun Paradox: Market-Oriented Production of Cloth in Eastern Canada in the Nineteenth Century." *Agricultural History* 76, no. 1 (2002): 28–57.

de Carufel, Hélène. "Le Lin." Ottawa: Leméac: 1980.

Dignam, M.E. "Weaving in Canadian Homes." *The Canadian Courier* 1, no. 13 (1907): 10–11.

Fair, Ross. "A Most Favorable Soil and Climate: Hemp Cultivation in Upper Canada, 1800–1813." *Ontario History* 96, no. 1 (2004): 41–61.

Faragher, John Mack. *A Great and Noble Scheme: The Tragic Story of the Expulsion of the French Acadians from Their American Homeland.* New York: W.W. Norton & Co., 2005.

Field, Richard Henning. "Lunenburg-German Household Textiles: The Evidence from Lunenburg County Estate Inventories, 1780–1830." *Material History Bulletin* 24 (1986): 16–23.

Gourlay, Robert. "Statistical Account of Upper Canada: Compiled with a View to a Grand System of Emigration." London: Simpkin & Marshall, 1822.

Grant, Janine, and Kris Inwood. "Gender and Organization in the Canadian Cloth Industry, 1870" in *Canadian Papers in Business History*, Volume I, edited by Peter Baskerville, pp. 17–31. Victoria, British Columbia: Public History Group, University of Victoria, 1989.

Greer, Allan. *Peasant, Lord, and Merchant: Rural Society in Three Quebec Parishes 1740–1840.* Toronto: University of Toronto Press, 1985.

Griffiths, N.E.S. *From Migrant to Acadian: A North American*

Border People, 1604–1755. Montreal: McGill-Queen's University Press, 2005.

_____. "The Golden Age: Acadian Life, 1713–1748." *Histoire Sociale/Social History* 17, no. 33 (1984): 21–34.

Gross, Laurence F. "Wool Carding: A Study of Skills and Technology." *Technology and Culture* 28, no. 4 (1987): 804-27.

Guillet, Edwin Clarence. *Pioneer Arts and Crafts.* Early Life in Upper Canada Series, Book 5. Toronto: Ontario Pub. Co., 1940.

Haight, Canniff. *Country Life in Canada Fifty Years Ago: Personal Recollections and Reminiscences of a Sexagenarian.* Toronto: Hunter, Rose, 1885.

Hood, Adrienne D. *The Weaver's Craft: Cloth, Commerce, and Industry in Early Pennsylvania.* Early American Studies. Philadelphia: University of Pennsylvania Press, 2003.

Hood, Adrienne D., and David-Thiery Ruddel. "Artifacts and Documents in the History of Quebec Textiles," in *Living in a Material World: Canadian and American Approaches to Material Culture,* edited by Gerald L. Pocius, pp. 55–91. St. John's, Newfoundland: Institute of Social and Economic Research, 1991.

Inwood, Kris, and Phyllis Wagg. "The Survival of Handloom Weaving in Rural Canada Circa 1870." *Journal of Economic History* 53, no. 2 (1993): 346–58.

"Journal and Proceedings of the House of Assembly, 1827." Halifax, N.S., s.n., 1827.

Kalman, Harold D. *A Textile Museum for Almonte: Phase 1: The Feasibility of Establishing a Textile Museum,* Ottawa: National Capital Commission, 1983. (Printed courtesy of the Commission.)

Lamontagne, Sophie-Laurence, and Fernand Harvey. "De l'économie familiale à l'artisanat: Les textiles domestiques." *Cap-Aux-Diamants* 50 (1997): 20–24.

_____. *La production textile domestique au Québec, 1827-1941: Une approche quantitative et régionale.* Collection Transformation. Ottawa: Musée national des sciences et de la technologie, 1997.

Mackley, Florence MacDonald. *Handweaving in Cape Breton.* Sydney, N.S.: privately printed, 1967.

McCalla, Douglas. "Textile Purchases by Some Ordinary Upper Canadians, 1808–1861." *Material History Review* 53 (Spring-Summer 2001): 4–27.

McCullough, A. B. The Primary Textile Industry in Canada: History and Heritage. *Studies in Archaeology, Architecture and History.* Ottawa: National Historic Sites, Parks Service, Environment Canada, 1992.

McKendry, Ruth McLeod. *Quilts and Other Bed Coverings in the Canadian Tradition.* Toronto: Van Nostrand Reinhold, 1979.

Montgomery, Florence M. *Textiles in America, 1650–1870.* New York: W.W. Norton Company, 1984.

Petersen, James B. *A Most Indispensable Art: Native Fiber Industries from Eastern North America.* 1st ed. Knoxville: University of Tennessee Press, 1996.

"Report of the Commissioner of the North-West Mounted Police 1890," in Sessional Papers of the Dominion of Canada: volume 15, first session of the seventh Parliament, session 1891, Ottawa: B Chamberlin, 1891.

Robertson, Margaret M. *Shenac's Work at Home: A Story of Canadian Life.* London: Religious Tract Society, 1868.

Roelens, Janine, and Kris Inwood. "'Labouring at the Loom': A Case Study of Rural Manufacturing in Leeds County, Ontario, 1870." *Canadian Papers in Rural History,* Volume VII, edited by Donald H. Akenson, pp. 215–35. Gananoque, Ontario: Langdale Press, 1989.

Ruddel, D.-T. "Consumer Trends, Clothing, Textiles and Equipment in the Montreal Area, 1792-1835." *Material History Bulletin* 32 (1990): 45-64.

_____. "The Domestic Textile Industry in the Region and City of Quebec, 1792-1835." edited by Gaston Tisdel, Robert D. Watt, and D. T. Ruddel. *Material History Bulletin* 17 (1983): 95-125.

Rygiel, Judith. "'Thread in Her Hands — Cash in Her Pockets': Women and Domestic Textile Production in 19th-Century New Brunswick." *Acadiensis* 30, no. 2 (2001): 56–70.

Strickland, Samuel, and Agnes Strickland. *Twenty-Seven Years in Canada West, or, The Experience of an Early Settler.* London: R. Bentley, 1853.

Traill, Catherine Parr. *The Backwoods of Canada: Being Letters from the Wife of an Emigrant Officer, Illustrative of the Domestic Economy of British America.* London: C. Knight, 1836.

_____. *The Canadian Emigrant Housekeeper's Guide.* Toronto: Lovell & Gibson, 1862.

Wallace-Casey, Cynthia. "'Providential Openings': The Women Weavers of Nineteenth-Century Queen's County, New Brunswick." *Material History Review* 46 (1997): 29–44.

Photo Credits

The publisher wishes to thank the curators and interpretive staff at the participating sites for supplying photographs from their own collections and for their co-operation and help during the photo shoots.

The images on the following pages were photographed by Rob Skeoch at Black Creek Pioneer Village, Toronto and Region Conservation Authority: 6, 11, 13 bottom, 26, 29, 31, 36, 37 upper L, 43 bottom, 46, 47, 48, 49, 50, 51, 54, 55L, 55R, 57, 70, back cover centre

The images on the following pages were photographed by Jackie MacRae of Behold Photographics at Upper Canada Village Heritage Park: front cover, 10, 22, 25, 28, 32, 33, 34, 37 bottom, 38, 39, 41L, 41R, 44L, 58 top, 59, 60, 61, 64, 66, 68, 69, 71, 72, 78

The image on the following page was supplied by the author: 65

Other photographs were supplied by and appear courtesy of:

Highland Village Museum: 8L, 8R, 27, 30, 37 upper R, 42, 62, 63, 74

Joseph Schneider Haus Museum and Gallery: 45, 56, 58 bottom, back cover bottom

Lang Pioneer Village: 9, 35

New Brunswick Department of Tourism and Parks: 12 top, 14

Ontario Science Centre: 52, 53

Peterborough Centennial Museum & Archives: 40, 44R

Textile Museum of Canada: 2, 75

The Ukrainian Cultural Heritage Village, Alberta Tourism, Parks, Recreation and Culture: 73, 76, 77

Village Historique Acadien: 3, 12 bottom, 13 top, 15, 16, 17, 19, 20, 21, 23, back cover top

R=right; L=left

Index

Acknowledgements

I would like to thank Lynn Schellenberg, Acquisitions Editor at James Lorimer & Company, for her multifaceted support of this project and especially for accommodating the unexpected need to extend some deadlines. Thanks too, to Holly Mazur, for invaluable assistance with the library research. Canada's textile history is a long, complicated story, and telling it here involved the cooperation of people from many historic sites across Canada who generously supplied a wealth of wonderful images from which to choose. Special thanks to Derek Cooke, Curator, Black Creek Pioneer Village, and Bruce Henbest, Supervisor of Mills, Trades and Heritage Programs, Upper Canada Village, for accommodating photo shoots at their sites, and to Susan Burke, Manager and Curator at Joseph Schneider Haus Museum, for personally photographing objects from her collection within very tight time constraints. I enjoyed working with photographers Jackie MacRae and Rob Skeoch — their photographs provide a fresh look at the Upper Canada Village and Black Creek sites. Finally, my most sincere thanks to all the researchers who work so hard to present Canada's early history to the public, and to the interpreters who make it come alive, both at their sites and in this book. This is dedicated to them.